Engineering drawing:

with CAD applications

Professor Dr Oleh Ostrowsky

Routledge
Taylor & Francis Group

LONDON AND NEW YORK

Preface

This book has been entirely updated to comply with the current BS 308. It gives an indepth coverage of engineering drawing and computer aided drawing objectives of all relevant Technician Education Council standard units at every level.

This new compact book replaces and consists mainly of material extracted from the previous *Engineering drawing for technicians*, volumes 1 and 2 by the same author. Both of these volumes were recommended for many engineering drawing courses in industry, schools colleges, polytechnics and universities. For instance, some polytechnics and universities were using volumes 1 and 2 as compulsory text for first year students on mechanical and production degree courses.

In order to make this book a comprehensive self-contained teaching unit on engineering drawing, some of the text has been rewritten and additional topics have been included. These consist of sections on electrical, electronic and fluid power symbols and diagrams, construction geometry and computer aided drawing, and the book completely covers the requirements of all relevant BTEC units on CAD.

This latest edition includes a new chapter on Introduction to Engineering Design, covering both conceptual and practical approaches to engineering design. The reader will also be aware of examples of practical design which appear on several pages of this book.

Chapters on loci, cams, bearings and basic engineering materials introduce students gradually to basic engineering design.

The contents of this book make it suitable of individuals who wish to teach themselves engineering drawing and designs on a distance learning basis. Lecturers engaged in teaching engineering drawing will also find it particularly useful, as all the information, explanations and sets of graduated examples needed for lecturer material are included.

When using this book as a workbook and in order to reduce the time consuming transfer of measurements, when preparing certain solutions also for testing purposes, 45 g/m2 office paper may be used as an economical substitute for tracing paper.

More than 280 test questions are included, with over 400 drawing examples suitable for classwork, homework, testing and examination purposes.

For manufacturing purposes, some examples incorporate construction squares until orthodox dimensioning is explained in more detail.

The relevant parts of BS 308 and BS 4500A are reproduced by kind permission of the British Standards Institution, 2 Park St., London W1A 2BS, from whom copies of the standards may be obtained.

Finally, for simplicity some drawing office personnel are described by the usual term 'draughtsmen and draughtwomen' or draughtspersons'.

O. Ostrowsky

First edition by Arnold 1989

Second edition published by Butterworth-Heinemann

This edition published 2013 by Routledge
2 Park Square, Milton Park, Abingdon, Oxon OX14 4RN
711 Third Avenue, New York, NY, 10017, USA

Routledge is an imprint of the Taylor & Francis Group, an informa business

British Library Cataloguing in Publication Data
A catalogue record for this book is available from the British Library

Library of Congress Cataloging-in-Publication Data
A catalog record for this book is available from the Library of Congress

ISBN: 978-0-415-50290-0

Contents

1 Communication

Throughout the ages, people have found communication with each other to be essential to their development. The means they have used have progressed from grunts to articulate speech and from signs and primitive drawings to competent writing and complicated drawings. All these have served to convey ideas, information, and instructions from one person to another.

In present-day industry, the principal means of communication is *engineering drawing*, which is the international language of engineering.

Engineering drawing is a system of communication in which ideas are expressed exactly, information is conveyed completely and unambiguously, and even the most complicated shapes are specifically described.

In Great Britain, the international conventions of engineering drawing are published by the British Standards Institution in British Standard BS 308, 'Engineering drawing practice'. This standard enables the draughtsman to understand clearly the designer's ideas and instructions and the craftsman to interpret precisely an engineering drawing for manufacturing or assembly purposes.

1.1 Engineering drawings

Engineering drawings are two-dimensional visual representations of three-dimensional objects and are used as a universal means of communication in industry.

Such drawings must be clear, concise, and accurate. They should convey, when required,
(a) information about the shapes, sizes and position of components,
(b) material requirements, and
(c) instructions about the method of manufacture.
All information must be complete and specified once only.

In general a formal engineering drawing may consist of three main parts:
(a) one or more views of an engineering component (object) or an assembly of components,
(b) dimensions, symbols, explanatory and instruction notes, and
(c) a title block.

Title block (p. 78, Table 7.2)
So that any drawing may be stored and, when required, identified and located quickly, an efficient system of labelling and cross-referencing is required. To facilitate this, all drawings must have a title block, which should contain the information required for identification and interpretation of the drawing.

Arrangements and positioning of title blocks differ considerably, but many drawing offices use sheets which are bought already printed and are of a standard size and layout.

1.2 Layout of drawings

Drawing sheets
Figure 1.1 shows the 'A' series of drawing sheets which are normally used:

    A4   210 mm × 297 mm
    A3   297 mm × 420 mm
    A2   420 mm × 594 mm
    A1   594 mm × 841 mm
    A0   841 mm × 1189 mm

The sides of all sheets are in the ratio $1:\sqrt{2}$. Area of the large A0 size is approximately $1\ m^2$.

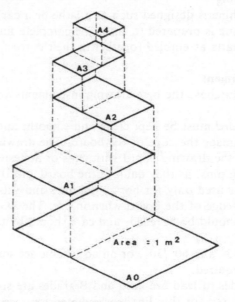

Fig. 1.1 The 'A' series of drawing sheets

1.3 Main types of drawing

Design layout drawing
Design layout drawings are usually sketches which represent feasible solutions to design problems and should include sufficient information for draughtsmen to produce the formal drawings (p. 90, Fig. 7.20).

Detail drawing
Detail drawings portray a single engineering component (object) and should contain sufficient information required to define that component completely in order to manufacture it (p. 63, Fig. 6.21 (a)).

Assembly drawing
Assembly drawings show the arrangement of several adjacent parts which together form a part of the finished assembled product, such as tailstock of a lathe or an engine of a car (p. 63, 6.21 (b) and p. 79, Fig. 7.1).

Combined drawing
This combination shows an assembly drawing and the detail drawings of constituent parts drawn on the same drawing sheet thus forming one drawing.

Arrangement drawing
When a large machine is designed such as a lathe or a car, a general arrangement drawing is prepared to show the complete finished product, with all parts assembled together in their correct positions.

1.4 Drawing equipment
It is advisable to purchase the best drawing instruments you can afford.

The *drawing board* must be kept clean and smooth, and care should be taken not to damage the edge of the board. The drawing paper should be fixed to the drawing board with clips or adhesive tape – never with drawing pins, as they damage the board and the paper.

The *tee square* is used only for horizontal lines and should be held tightly against the edge of the board when in use. The working edge of the tee square should be bevelled, and care should be taken not to damage it.

Set squares of 45° and 60°/30°, or an adjustable set square with bevel edges, are required.

Pencils H grades of lead are hard and B grades are soft. The 2H grade is generally used for thin line work, dimensions, centre lines,

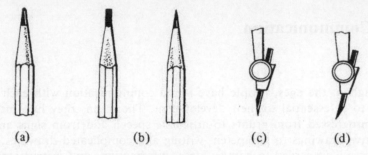

Fig. 1.2   (a) Pencil sharpened to a cone point
                (b) Two views of a pencil sharpened to a chisel point
                (c) Small spring-bow-compass lead sharpened in one plane on the inside
                (d) Large spring-bow-compass lead sharpened in one plane on the outside

hidden detail, etc. The H grade is used for thick line work, visible outlines etc. The HB grade is used for lettering, numerals, and sketching.

Pencils can be sharpened into a cone point or a chisel point, using a sandpaper block, as shown in Fig. 1.2. The advantage of the chisel point is that thick and thin lines can be drawn with the same pencil, the edge is retained longer, and it is less likely to break. A cone will last longer if the pencil is rotated occasionally while drawing or lettering.

Compasses of the spring-bow type with a shouldered pin are preferable. Compass leads should be of the same grade as the drawing pencils used. They should be sharpened in one plane only – on the inside for small compasses and on the outside for large compasses, as shown in Fig. 1.2 (c) and (d). The different grades of leads required could be taken from unwanted drawing pencils.

It is desirable to use three sizes of compass: a small spring-bow, a large spring-bow, and a beam compass for very large arcs.

Compasses may be used as *dividers*, if required, by replacing the leads by pins.

Scales should be marked accurately in divisions of 1 mm or, preferably, 0·5 mm over the full length.

A *radius curve* is useful for drawing internal and external fillet radii.

A *French curve* is very useful for drawing curves other than circular curves.

The *eraser* should be a soft white pencil rubber, to ensure that the drawing-paper surface will not be damaged.

An *eraser shield* is very useful for erasing mistakes on drawings without erasing adjacent correct lines.

A *sandpaper block* or a small smooth file is used for sharpening leads.

A *clean duster* is useful for keeping drawing equipment clean.

Hands must always be clean when drawing.

Table 1.1 Types of line

Type		Description	Application
A	————————	Thick continuous	Visible outlines and edges
B	————————	Thin continuous	Dimensions and leader lines, projection lines, hatching lines, short centre lines, and revolved sections
C₁	∿∿∿∿	Thin continuous irregular or	Limits of partial or interrupted views or sections when the line is not an axis
C₂	─/\─/\─	straight with zig-zag	
D	– – – – –	Thin or thick short dashes	Hidden outlines and edges
E	—·—·—·—	Thin chain	Centre lines, lines of symmetry, pitch circles and lines
F	—·—·——	Chain, thick at ends and at changes of direction, thin elsewhere	Cutting plane
G	—··—··—	Thin chain short double dashes	Outlines and edges of adjacent parts and extreme positions of moveable parts.

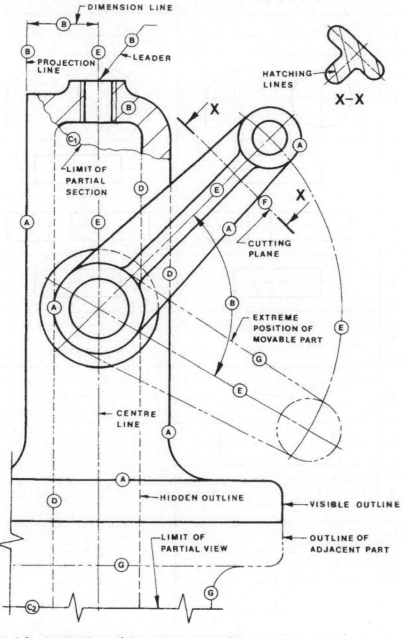

Fig. 1.3 Applications of the various types of line

3

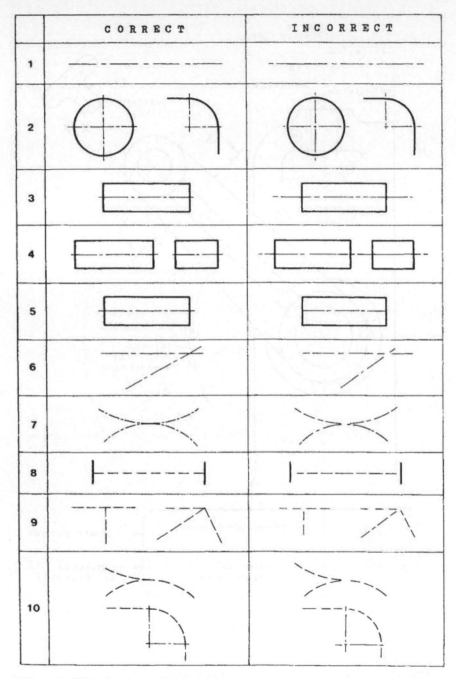

Fig. 1.4 General rules on line work

1.5 Lines, letters and numerals
All lines should be dark, bold, and of consistent density and thickness.

Two thicknesses of line are recommended: a thin line and a thick line two to three times thicker than the thin line (see p. 3).

When two or more lines of different type coincide, a visible outline (type A) *always takes precedence over any other type of line* (p. 30, Fig. 5.10, No. 7), *and a dashed thin line showing hidden detail* (type D) *is drawn in preference to a centre line* (type E), (p. 30, Fig. 5.10, No. 8).

General rules on line work (see Fig. 1.4)
1 All chain lines should start and finish with a long dash.
2 Centre lines should cross one another at long-dash portions of the line at the centres of circles, arcs, etc.
3 Centre lines should extend only a short distance beyond the feature, unless required for dimensioning etc.
4 A centre line should not extend through the spaces between views.
5 A centre line should not terminate at another line.
6 Where angles are formed in chain lines, long dashes should meet or cross at the intersections.
7 Arcs should join at tangent points.
8 Dashed lines should start and end with dashes in contact with the hidden or visible outline.
9 Dashed lines should meet or cross with dashes at the intersection.
10 If a dashed line meets a curved line tangentially, it should be with solid portions of the lines.

Lettering and numerals

ABCDEFGHIJKLMNOPQRSTUVWXYZ
abcdefghijklmnopqrstuvwxyz
1234567890

ABCDEFGHIJKLMNOPQRSTUVWXYZ
abcdefghijklmnopqrstuvwxyz
1234567890

1 Only one chosen style should be used. Vertical and sloping letters should not be mixed on one drawing.
2 Uppercase (capital) letters are preferred to lowercase (small) letters.
3 Height of lowercase letters should be 0.6 times that of uppercase.
4 There should be equal spacing between letters of about one-third of a letter width, and the spacing between words not less than a letter width.

5 Spacing between lines of notes should not be less than one-half of the letter height.

6 Dimensions and notes should be not less than 2.5 mm tall, but titles and drawing numbers should not be less than 7 mm high.

7 All notes should be placed in horizontal position.

8 Underlining of notes is not recommended – larger letters may be used.

1.6 Technical terms

It is necessary for engineers to know and understand the technical terms describing components and their features.

1 A *flange* is a projecting thin disc on pipes or couplings joining two shaft ends together.

2 and 3 A *keyway* is a groove in a shaft or a hub machined to accommodate a corresponding key.

4 A *key* is a piece of shaped metal which is inserted in a shaft and a hub to prevent relative movement between those two parts.

5 A *taper* is a gradual change in diameter of a component along its length.

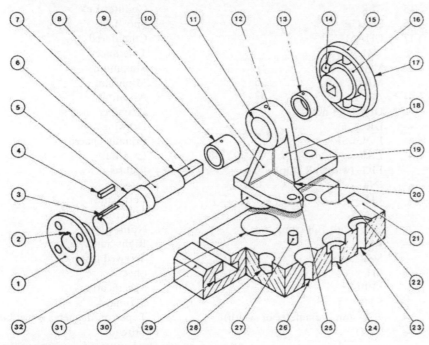

Fig. 1.5 Technical Terms

6 A *shaft* is a cylindrical rotating rod upon which parts are fixed, used for transmission of motion.

7 A *shoulder* is a sudden change in diameter.

8 A *square on a shaft* is a length of the shaft with a square cross-section.

9 A *bush* is a plain bearing supporting a rotating shaft and can easily be replaced when worn out.

10 A *web* is a thin flat part connecting heavier parts of a component, it is usually *parallel* to the bosses, bores, shafts, etc.

11 A *bore* is a cylindrical hole along a tube or a boss.

12 A *boss* is an enlarged protruding round part of a casting, used to accommodate a hole.

13 A *collar* is a separate ring of rectangular section or an integral part of a shaft used for axial location.

14 *Spokes* are rods radiating from the hub to the rim of a wheel.

15 A *rim* is the outer part of a wheel.

16 A *hub* is the inner part of a wheel.

17 A *pulley* is a small wheel with a flat or grooved rim to carry a belt, rope, etc.

18 A *rib* is a thin part used to support or strengthen heavier parts of component, it is usually *perpendicular* to the bosses.

19 A *bracket base* is the bottom part of a projecting support, usually fixed to a flat surface.

20 A *fillet* is an internal corner of a casting etc. which is curved to assist the flow of molten metal during casting and also to make the corner stronger by reducing stress concentrations.

21 A *table* is the flat top on which working components can be fixed.

22 A *slot* is an elongated hole or groove.

23 A *spot-faced* surface is a flat circular surface concentric with a hole, used for seating screw heads etc.

24 A *counterbored hole* is a hole, part of which is of larger diameter and flat-bottomed to conceal screw heads etc.

25 A *lug* is a projection from a casting etc., used for fastening and adjusting purposes.

26 A *countersunk hole* is a hole, part of which is conical to receive screw heads.

27 A *dowel* is a headless cylindrical pin used for precise-location purposes.

28 A *blind-drilled hole* is a hole which does not pass completely through the component.

29 A *tee groove* or *tee slot* is a long aperture used to accommodate fixing bolts, preventing them turning.

30 A *chamfer* is a surface produced by bevelling square edges.

31 A *recess* is a shallow hole to suit the shape of a spigot or a similar matching part.

32 A *spigot* is a projection which fits into a corresponding recess and is used for precise-location purposes.

1.7 Abbreviations and symbols

There are a number of common engineering terms and expressions which are frequently replaced by abbreviations or symbols on drawings, to save space and draughting time. Some of the abbreviations and symbols recommended by the British Standards Institution in BS 308 are illustrated below and listed on the right.

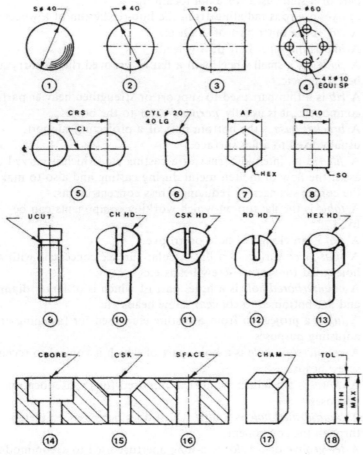

Fig. 1.6 Abbreviations and symbols

Table 1.2 List of abbreviations and symbols

No.	Abbreviations or symbol	Term
1	S∅ (preceding a dimension)	Spherical diameter
2	∅ (preceding a dimension)	Diameter
3	R (preceding a dimension)	Radius
4	PCD	Pitch-circle diameter
5	CRS	Centres
	₵ (on a view)CL(in a note)	Centre line
6	CYL	Cylinder or cylindrical
	LG	Long
7	AF	Across flats
	HEX	Hexagon or hexagonal
8	□ (preceding a dimension) or ⊠	Square
	SQ (in a note)	Square
9	UCUT	Undercut
10	CH HD	Cheese head
11	CSK HD	Countersunk head
12	RD HD	Round head
13	HEX HD	Hexagonal head
14	CBORE	Counterbore
15	CSK	Countersunk
16	SFACE	Spot face
17	CHAM	Chamfered
18	TOL	Tolerance
	MAX	Maximum
	MIN	Minimum
—	ASSY	Assembly
—	DIA (in a note)	Diameter
—	DRG	Drawing
—	EQUI SP	Equally spaced
—	EXT	External
—	FIG. (with full stop)	Figure
—	INT	Internal
—	LH	Left-hand
—	MATL	Material
—	NO. (with full stop)	Number
—	RH	Right-hand
—	SCR	Screwed or screw
—	SH	Sheet
—	SPEC	Specification
—	STD	Standard
—	▷ (on a diameter or width)	Taper on diameter or width
—	THD	Thread

1.8 Conventional representation of common features

There are many common engineering features which are difficult and tedious to draw in full. In order to save draughting time and space on drawings, these features are represented in a simple conventional form as shown in Figs 1.7–1.9.

(a) *External screw threads*

The crests on the male thread of a stud are defined by a continuous thick line, and the roots of threads by a parallel continuous thin line. The distance between these parallel lines should be approximately equal to the depth of thread, i.e. approximately one tenth of the major diameter of the thread (see p. 98, 106 and 107).

The limit of the useful length of the thread – 'full thread' – is shown by a continuous thick line (see p. 106, Fig. 10.3 and p. 107, Fig. 10.5).

Incomplete threads, run-outs beyond the limits of useful thread length are not normally shown unless there is a special functional need.

In an end view, the thread roots inside the material are represented by an inner thin broken circle.

(b) *Internal screw threads*

The tapped hole initially is drilled, which is indicated by the thick outlines. When the hole is tapped, the roots of the threads are defined by a parallel continuous thin line.

In an end view, the thread roots inside the material are represented by an outer thin broken circle.

In a sectional view, the hatching lines are drawn across the thin lines.

(c) *A screw-thread assembly*

The male thread of an inserted stud takes precedence over the female thread of the hole.

The hatching lines are not drawn across the thick lines.

In an end view, the male part which is nearest to the observer is represented.

(d) *Interrupted views*

To save space, it is permissible to show only those parts of a long component which are sufficient for its definition. All break lines are thin and continuous.

(e) *Flat features on a shaft*

To avoid drawing an additional view, a square, tapered square or a local flat on a round part, may be indicated by two diagonal continuous thin lines.

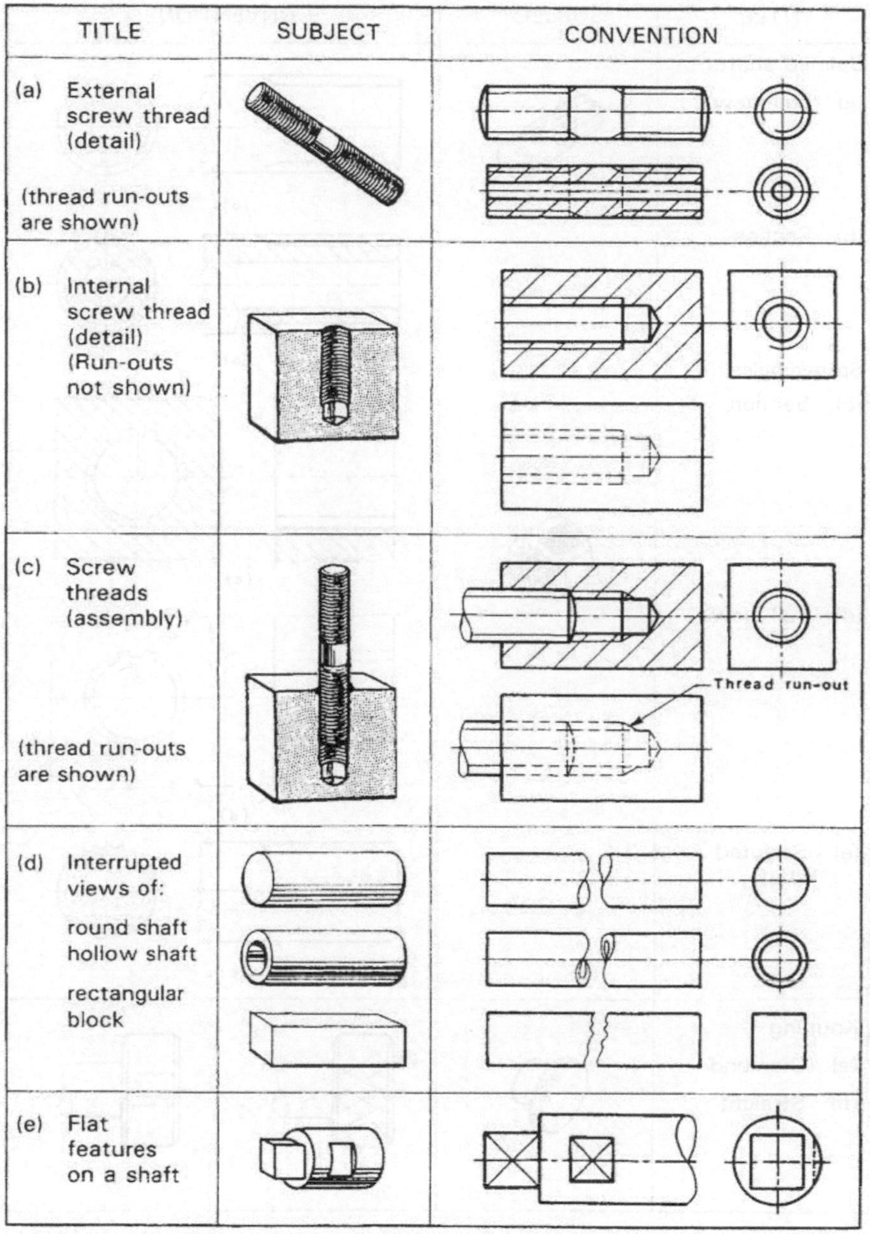

TITLE	SUBJECT	CONVENTION
(a) External screw thread (detail) (thread run-outs are shown)		
(b) Internal screw thread (detail) (Run-outs not shown)		
(c) Screw threads (assembly) (thread run-outs are shown)		Thread run-out
(d) Interrupted views of: round shaft hollow shaft rectangular block		
(e) Flat features on a shaft		

Fig. 1.7 Conventional representation of common features

7

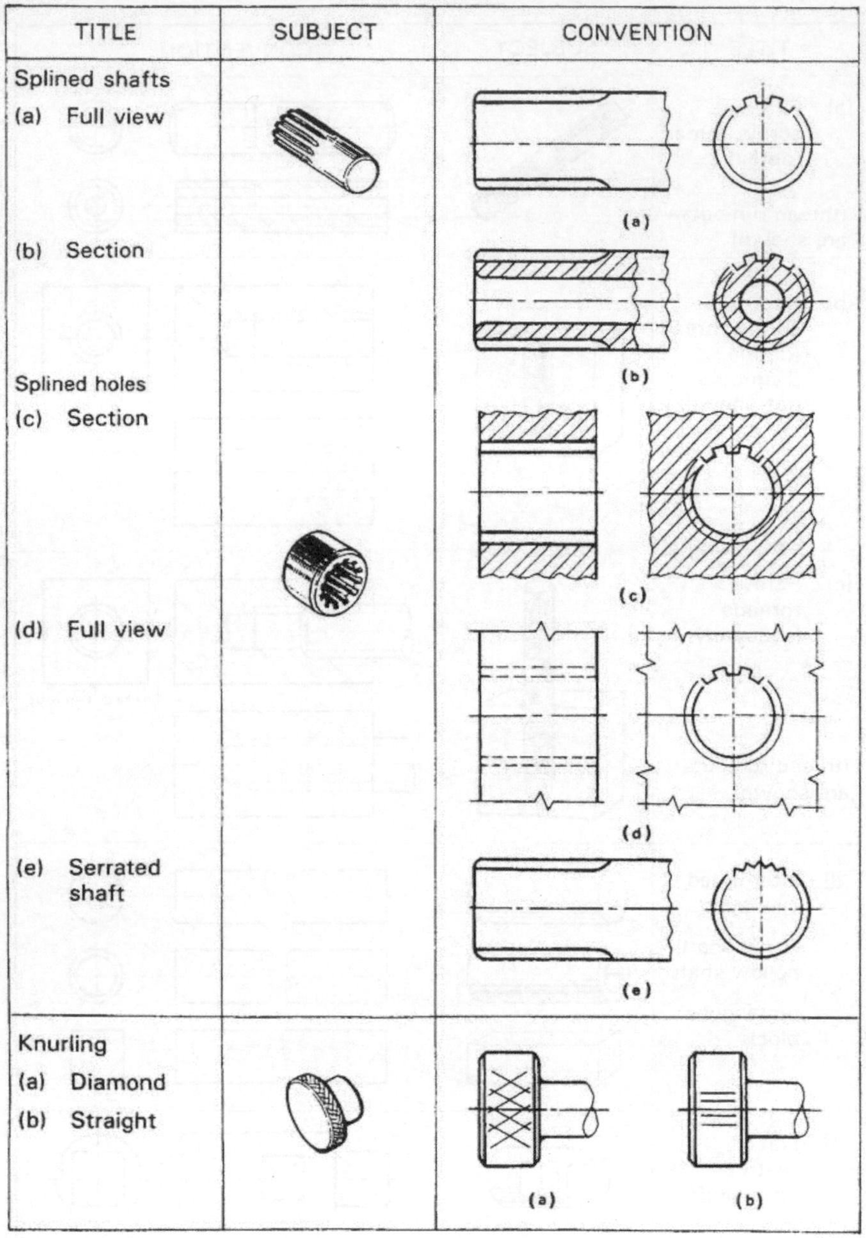

TITLE	SUBJECT	CONVENTION
Splined shafts (a) Full view		 (a)
(b) Section		 (b)
Splined holes (c) Section		 (c)
(d) Full view		 (d)
(e) Serrated shaft		 (e)
Knurling (a) Diamond (b) Straight		 (a) (b)

Fig. 1.8 Conventional representation of the other common features

Splines and serrations

Splines are a number of integral keys produced by machining longitudinal grooves in a shaft. Similarly the grooved ways are formed in a hole of another component into which the splined shaft is to fit.

(a) In the longitudinal axial view the roots of splines are represented by thick lines whereas in comparison the roots of threads are represented by thin lines.

In the end view only a few splines need be shown, with the root circle represented by an inner thin circle.

(b) The cross sectional view is basically similar to the outside view except for cross hatching.

(c) The splined hole root circle is represented by an outer thin circle. In the longitudinal axial view the roots of splines are represented by thick lines. The hatching lines are not drawn across these thick lines, whereas in comparison the hatching lines are drawn across the thin lines representing the roots of threads.

(d) The partial views are drawn using a thin overlapping line with zig-zags where the partial views in (c) are drawn using a continuous thin irregular line.

(e) The same conventions apply for serrations as for splines.

Diamond and straight knurling

Knurling provides a rough surface to facilitate the operation of the component by hand. (A spring-bow compass usually incorporates both types of knurling).

Cylindrical helical springs

(a) A compression spring of wire of circular cross section;

(b) A compression spring of wire of rectangular cross section;

(c) A tension spring of wire of circular cross section;

Coils are drawn only at each end of the spring and are connected by centre lines indicating the repetition of coils.

The springs may be drawn diagramatically as shown in simplified representation. The wire section may be indicated by symbol, ▢ for square and ϕ for round wire.

Spur gears

(a) The front view of a spear gear consists of the outside thick circle and the pitch 'centre-line' circle.

In the sectional end view the gear teeth are not sectioned, but the centre line and the thick line representing the root of the teeth are shown.

In the remaining end views the centre lines are shown and the direction of teeth is indicated by thin parallel lines.

(b) In the assembly drawing neither of two gears in mesh is assumed to be hidden by the other. The direction of the teeth should be shown on one gear only.

The conventional representations apply irrespective of the type of gear teeth or number of teeth.

Repetitive features

Repeated drawings of identical features may be avoided by drawing only one feature and indicating the position of others by centre lines.

Continuous thin lines are used for drawing the short centre lines.

Bearings

Sometimes it is necessary to draw sectional views of a number of ball and roller bearings on one drawing. The complicated sectional views can be replaced by a conventional representation consisting of the bearing outline and thin-line diagonals. This conventional representation does not imply any particular type or detail of bearing.

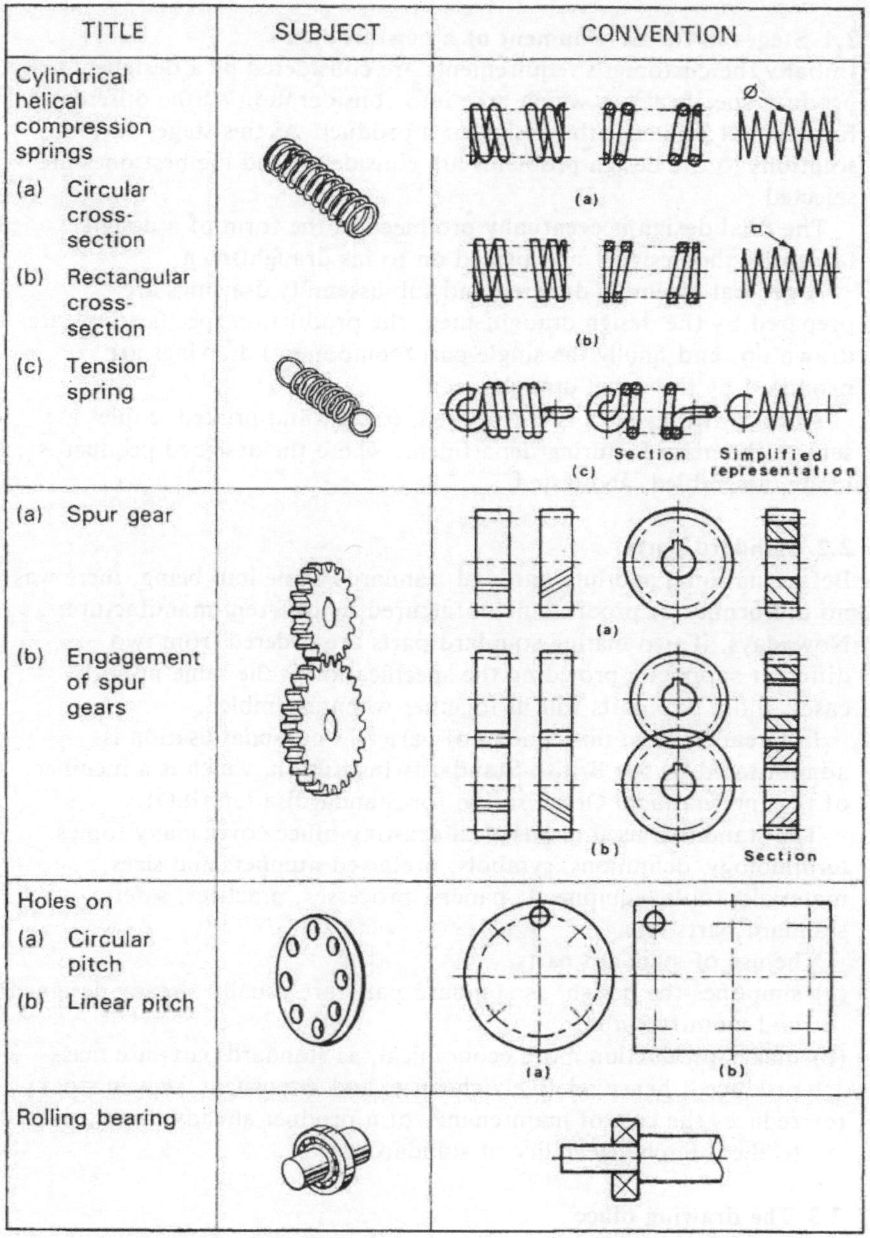

TITLE	SUBJECT	CONVENTION
Cylindrical helical compression springs (a) Circular cross-section (b) Rectangular cross-section (c) Tension spring		(a) (b) (c) Section Simplified representation
(a) Spur gear (b) Engagement of spur gears		(a) (b) Section
Holes on (a) Circular pitch (b) Linear pitch		(a) (b)
Rolling bearing		

Fig. 1.9 Further common features

9

2 Organisation

2.1 Stages in the development of a new product

Initially the customer's requirements are considered by a designer, to produce specifications which take into consideration all the different factors that influence the design of a product. At this stage, various solutions to the design problems are considered and the best ones are selected.

The final design is eventually produced in the form of a design layout by the designer and passed on to his draughtsmen.

A general-assembly drawing and sub-assembly drawings are prepared by the design draughtsmen, the production specifications are drawn up, and finally the single-part (component) drawings are produced by the detail draughtsmen.

After drawings have been checked, traced, and printed, copies are sent to the manufacturing department, where the designed product is made, assembled, and tested.

2.2 Standard parts

Before national and international standards came into being, there was no uniformity of products manufactured by different manufacturers. Nowadays, if two mating standard parts are ordered from two different suppliers, providing the specification is the same in both cases, those two parts will fit together when assembled.

In Great Britain, this control of variety, or standardisation is administered by the British Standards Institution, which is a member of the International Organisation for Standardisation (ISO).

The standards used in a typical drawing office cover many topics: terminology, definitions, symbols, preferred numbers and sizes, materials, tools, equipment, papers, processes, practices, safety, standard parts, etc.

The use of standard parts
(a) simplifies the design, as standard parts are usually already designed and manufactured;
(b) makes production more economical, as standard parts are mass-produced, hence relatively cheaper, and are usually kept in stock;
(c) reduces the cost of maintenance of a product already in use, due to the interchangeability of standard parts.

2.3 The drawing office

The functions of a drawing office vary from firm to firm. A relatively large firm may have a separate design office, whereas the majority of firms have drawing offices incorporating the design section.

The main functions of a typical drawing office are
(a) to prepare the design layouts and assembly and production drawings necessary for the manufacture of products;
(b) to make decisions on materials to be used, methods of manufacture, heat treatment, etc.;
(c) to calculate the stresses for the designed components, to ensure that the components will withstand the applied forces when manufactured and in use;
(d) to estimate from the drawings the cost of manufacture of designed components;
(e) to store all drawings, technical information, and reference material;
(f) to provide a technical service for all departments in the firm; and
(g) to liaise with people outside the firm.

2.4 The print room (reprographics)

The function of a reproduction or print room is to copy engineering drawings, tracings, and documents. The copies obtained are used for reference, manufacturing, assembly, and storage purposes. There are four popular reproduction methods: the dye-line (diazo) process, the microfilm process, the xerographic process, and computer plotting (see p. 120).

The dye-line (diazo) process

The tracing is placed over special printing paper and is exposed to ultra-violet light, which bleaches away the sensitive coating on the printing paper, except where the ink lines on the tracing prevent light passing. The exposed paper is then developed, to show the lines of the tracing in dark colours.

The print obtained is the same size as the original drawing.

The microfilm process

The original drawing or tracing is photographically reduced on to a film. The film, when developed, may be stored on a reel or may be cut and placed in cellophane envelopes or mounted in cards or frames. When the drawing is to be referred to, the film is projected on to a screen or enlarged and printed directly.

The original drawing or tracing must be of good standard – the lines must be drawn black and thick and well spaced, detail reduced, and lettering clear.

Xerographic process

This uses a plate coated with a material which conducts electricity when exposed to light and acts as an insulator in the dark. The plate surface is positively charged in the dark and a drawing is projected on to it. Plate areas corresponding to white areas of the drawing are exposed to light and lose their charge; dark areas, corresponding to the image of the drawing, retain their positive charge.

The plate is then dusted with a negatively charged powder which adheres only to the positively charged areas. A positively charged paper is pressed against the plate, attracting the particles of powder, and finally heat is applied to fuse those particles permanently to the paper.

This process is suitable only for reproducing small-sized drawings and documents.

2.5 Drawing-office personnel

Designer

A designer is usually a professional engineer, a technologist with a degree. He or she must be creative – be able to develop ideas and solve design problems. A designer's technical knowledge should include such disciplines as mathematics, theory of machines, strength of materials, fluid mechanics, thermodynamics, materials science, production methods, electricity, electronics, ergonomics, aesthetics, etc.

He or she must be able to convey ideas and instructions clearly, accurately, and concisely, to guide the draughtsmen.

The designer's main functions are

(a) to originate and finalise designs for a proposed product to satisfy the function, cost, manufacture, and market requirements in consultation with production engineers, industrial (artistic) designers, purchasing specialists, customers, and other interested persons;

(b) to advise the drawing-office staff on technical difficulties;

(c) to ensure that the product, after it has been manufactured, is functional, reliable, and easily maintained.

A designer may hold the post of chief draughtsman or drawing-office manager.

Chief draughtsman

The chief draughtsman is a professional or technician engineer with a higher technician diploma or certificate. His or her functions as drawing-office manager are

(a) to organise, direct, and co-ordinate the work of the drawing-office and its resources;

(b) to determine staffing, promotion, and necessary training for junior staff;

(c) to co-ordinate the activities of drawing-office staff – section

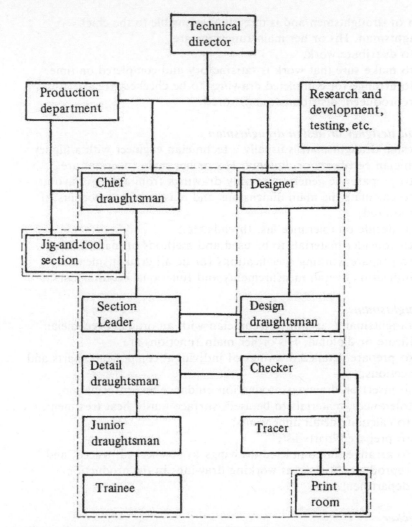

Fig. 2.1 Organisation of a typical drawing office and design office

leaders, design and detail draughtsmen, checkers, tracers, etc.;

(d) to advise staff on technical problems;

(e) to liaise with other departments inside the firm and with people outside the firm;

(f) to examine design specifications provided by a designer;

(g) to instruct section leaders on the distribution and detailing of work.

Section leader

The section leader is usually a technician engineer supervising a small

team of draughtsmen and is directly responsible to the chief draughtsman. His or her main functions are
(a) to distribute work,
(b) to make sure that work is satisfactory and completed on time,
(c) to arrange for completed drawings to be checked, traced, reproduced, distributed, and stored.

Detail designer or design draughtsman

A design draughtsman is usually a technician engineer with a higher technician certificate or diploma. His or her main functions are
(a) to prepare the general-assembly drawings from a design layout;
(b) to calculate the main dimensions and masses of components, if required;
(c) to decide on tolerance fits, threads etc.;
(d) to consider materials to be used and methods of manufacture;
(e) to prepare working specifications for detail draughtsmen, indicating overall measurements and functional requirements.

Draughtsman

A draughtsman is usually a technician with an ordinary technician certificate or diploma. His or her main functions are
(a) to prepare detailed drawings of individual components, parts and sections;
(b) to insert on drawings production guidance and information, tolerances, materials to be used, surface finish, heat treatment, etc.;
(c) to calculate detail dimensions;
(d) to prepare a parts list;
(e) to arrange for completed drawings to be checked, traced, and reproduced for use as working drawings in the production department.

Checker

The checker is usually a technician, and his or her main functions are
(a) to check finished engineering drawings to ensure that all specification requirements have been met, all dimensions and tolerances have been shown, and all information has been inserted regarding manufacture, materials to be used, surface finish, heat treatment, etc.;
(b) to check the accuracy of dimensions;
(c) to ensure that safety has been considered;
(d) to make sure that all parts will function correctly and are easy to manufacture.
Finally, the checker refers drawings back to the draughtsmen for ratification.

Tracer

The tracer's main functions are
(a) to make tracings, usually in ink, of the original drawings;
(b) to insert additional detail, if instructed;
(c) to refer drawings to the print room and then to file drawings, tracings, and copies of drawings if required.

Drawings are traced in order to improve the quality of prints and also to make a more durable record of original drawings.

Every draughtsman and tracer should know about the different methods of drawing reproduction, in order to get the best possible prints from his or her drawings.

2.6 Test questions on chapters 1 and 2

1 State what is meant by engineering drawing.
2 Give reasons for using a drawing in preference to written and spoken communication in engineering.
3 Drawings should convey three types of general information. Name two of them.
4 State the function of a single-part drawing and list the items it should include.
5 State the function of an assembly drawing.
6 Name five items of basic information that a drawing title block should include.
7 State why the following types of drawing are needed:
(a) single-part or detail,
(b) assembly,
(c) general arrangement.
8 If a thick continuous line is used to represent the visible outlines and edges of a component on a drawing, what will the following lines represent?
(a) a thin continuous line,
(b) thin short dashes,
(c) a thin chain line.
9 Which of the following statements are true and which are false?
(a) A visible outline takes preference over any other line on a drawing.
(b) A centre line takes preference over a thin dashed line showing hidden detail.
(c) Centre lines should cross one another at long-dash portions of the lines.
(d) Underlining of notes is not recommended.
10 With the help of simple sketches, identify and explain the following abbreviations or symbols.
   CRS AF UCUT HEX HD TOL LG
   SFACE CHAM □ PCD CL STD RH

11 Explain briefly why standard abbreviations are used on engineering drawings.

12 With the help of sketches, show the conventional representation of the following:
   (a) an external thread,
   (b) an interrupted view of a hollow shaft,
   (c) a square on a shaft,
   (d) diamond knurling,
   (e) a splined shaft,
   (f) a bearing.

13 Put the following stages in the development of a new product in their correct order, naming the personnel involved and their functions: (a) single-part drawing, (b) manufacture of a product, (c) printed copy of a drawing, (d) assembly of a product, (e) design layout, (f) assembly drawing.

14 State four main functions of a typical drawing office.

15 Indicate the reasons for copying engineering drawings.

16 Describe very briefly two popular drawing–reproduction processes.

17 Describe at least three main functions of each of the following drawing-office personnel:
   (a) the chief draughtsman,
   (b) the designer,
   (c) the draughtsman,
   (d) the checker,
   (e) the tracer.

18 State three reasons for using standard parts in engineering.

19 With the help of a simple block diagram indicate the organisation of a drawing office.

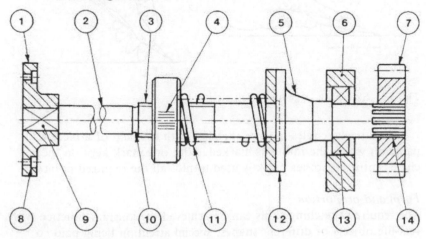

Fig. 2.2 Test question 20

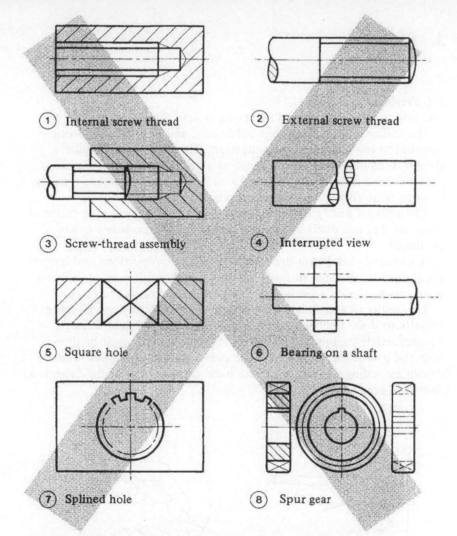

① Internal screw thread   ② External screw thread

③ Screw-thread assembly   ④ Interrupted view

⑤ Square hole   ⑥ Bearing on a shaft

⑦ Splined hole   ⑧ Spur gear

Fig. 2.3 *Incorrect* conventional representations

20 Study Fig.2.2 and identify the referenced items with the help of the following descriptions: (a) square on shaft, (b) counterbored hole, (c) interrupted view, (d) bearing, (e) external thread, (f) undercut, (g) straight knurling, (h) splined shaft, (i) compression spring, (j) taper, (k) bearing housing, (l) shaft coupling, (m) spur gear, (n) shaft flange.

21 Figure 2.3 shows conventional representations of common features, which are *not* drawn in accordance with BS 308. Redraw each item correcting all deliberate mistakes. Items 1, 2, 4, 5, 6 and 7 include at least three mistakes; items 3 and 8 include at least six mistakes. Tracing paper may be used.

3 Sketching

3.1 Freehand sketching

The importance of freehand sketching is very often underestimated.

The ability to sketch quickly, accurately, and in good proportion is essential to engineering communication. The freehand technique should be employed by an engineer as a better means of visualising problems and quickly organising his or her thoughts to avoid wasting time on more formal drawing methods.

The designer nearly always sketches his or her first ideas in pictorial form, as they are easily visualised before the proper drawings are produced.

A freehand sketch is a drawing in which all proportions and lengths are judged by eye and all lines are drawn without the use of drawing instruments—the only tools used are pencil, eraser, and paper.

The ability to sketch is a skill which is acquired through learning initially to draw freehand vertical and horizontal straight lines, squares, circles, ellipses, and curves. Circular curves must be drawn with the ball of the hand inside the curve, and straight lines must be drawn by resting the weight of the hand on the backs of the fingers, as shown in Fig. 3.1.

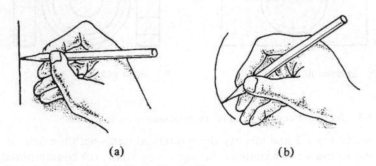

(a)   (b)

Fig. 3.1 Hand positions for sketching

To sketch a straight line (Fig. 3.1(a))

1 Mark the end points of the required line.
2 Sketch a light trial line using several short strokes, with the eye fixed on the point towards which the straight line is being drawn.
3 Finally, press the pencil down to get a uniform bold straight line.

To sketch a circle (Fig. 3.2)

1 Sketch centre lines and the enclosing construction square, then sketch the diagonals and step of distances from the centre equal to the radius.
2 With the hand positioned within the circle and pivoted at the wrist, sketch the trial circle consisting of eight short arcs.
3 Finally, press the pencil down to get a uniform bold line and erase all construction lines as required.

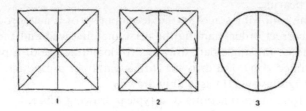

Fig. 3.2 Three stages in sketching a circle

Circles may also be drawn by rotating the paper with the left hand about one of the fingers of the right hand acting as a pivot, while the right hand holds the pencil at the required length. Figure 3.3 shows how to hold the pencil for sketching large and small circles.

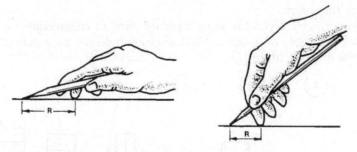

Fig. 3.3 Sketching large and small circles

Alternatively, circles may be sketched with the help of a piece of paper on which the radius is marked. With one mark kept on the circle centre, the other mark is used to plot all the required points.

Form and proportion

The required sketching skills can be achieved by constant practice with real-life objects of different shapes, special attention being paid to density of line, good form, and relative proportions.

To obtain good form and proportion, a light construction framework of rectangular boxes, cubes, cylinders, cones, etc. can be used to represent the outlines of the objects sketched, as shown in Fig. 3.4.

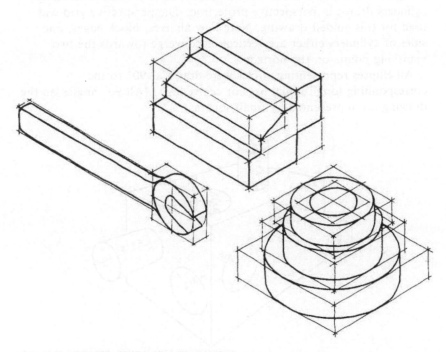

Fig. 3.4 Use of construction framework

To determine the required proportions, the pencil should be held at arm's length, marking the height of the object with the tip of the thumbnail, Fig. 3.5. The arm then is rotated until the pencil coincides with another edge of the object and then an estimate is made along the pencil of the ratio of the two lengths.

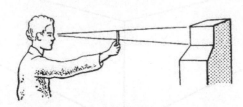

Fig. 3.5 Measuring proportions for sketching

4 Pictorial projection

Projection is a method of representing visually a three-dimensional object on two-dimensional drawing paper. A pictorial projection is a method of producing a two-dimensional view of a three-dimensional object that shows three main faces indicating the height, width, and depth simultaneously, as in Fig. 4.1.

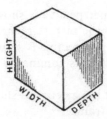

Fig. 4.1 Representation of a three-dimensional object

4.1 Perspective projection
Appreciation of perspective is essential for learning the fundamentals of freehand sketching. Objects at a distance appear to be smaller than those which are near. Two parallel lines representing the edges of a straight road seem to come closer together and then meet at a point on the horizon. That point is called the vanishing point (VP), Fig. 4.2.

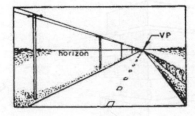

Fig. 4.2 Perspective projection

Perspective projection involves a number of receding lines called projectors converging at one, two, or more vanishing points. The objects sketched are then presented as they would appear when observed from a particular point in real life.

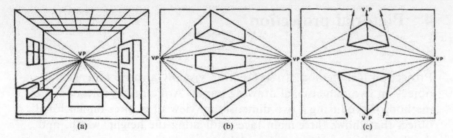

Fig. 4.3 (a) One-point, (b) two-point and (c) three-point perspective

In Fig. 4.3(a), one-point perspective is shown, where one of the principal faces is parallel to the picture plane.

In Fig. 4.3(b), two-point perspective is shown, where all principal faces are inclined. This method is commonly used for industrial sketching.

In Fig. 4.3(c), three-point perspective is shown, where one vanishing point is outside the picture frame.

For engineering purposes, when sketching small objects, the vanishing points are considered to be placed outside the frame of the drawing paper. This is because such objects are usually viewed from a close distance.

In Fig. 4.4, the features of the objects above the eye-line or horizon are seen from below, and the features below the horizon are seen from above.

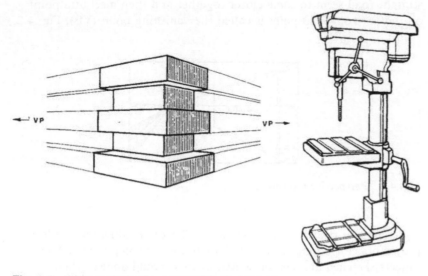

Fig. 4.4 Objects drawn in perspective projection

For guided perspective sketching, use the grid in Fig. 4.8. Place tracing paper over the grid and sketch the objects required using the projectors. Alternatively, draw your own projectors radiating from two vanishing points on the horizon.

Figure 4.5 shows a block with a number of holes and protruding cylinders drawn in perspective projection. The perspective grid was used for this guided drawing. Note how all axes, block edges, and sides of cylinders either are vertical or converge towards the two vanishing points on the horizon.

All ellipses representing circles were drawn at 90° to the corresponding longitudinal axes or centre lines. (All 90° angles on the drawing are represented by small squares).

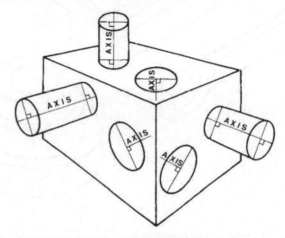

Fig. 4.5 Circular shapes in perspective projection

4.2 Test questions
1 The detail shown in Fig. 4.6 requires holes to be drilled at the positions marked + and cylindrical (dowel) pins to be inserted at the positions marked *.

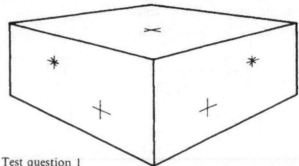

Fig. 4.6 Test question 1

Redraw or trace the block and then sketch the holes and pins. The holes are to be approximately 15 mm diameter and the dowels approximately 10 mm diameter, protruding 20 mm from the block. Hole and pin axes are to be normal (at 90°) to the block surfaces.

2   Figure 4.7 shows the outlines of tools that are used by engineering workers of different trades: (1) hammer, (2) double-ended spanner, (3) pliers (4) adjustable spanner, (5) brace, (6) G-clamp, (7) hand saw, (8) tool-maker's clamp, (9) blow lamp, (10) bench vice, and (11) micrometer.

Sketch freehand each tool in perspective projection, using the construction-box method.

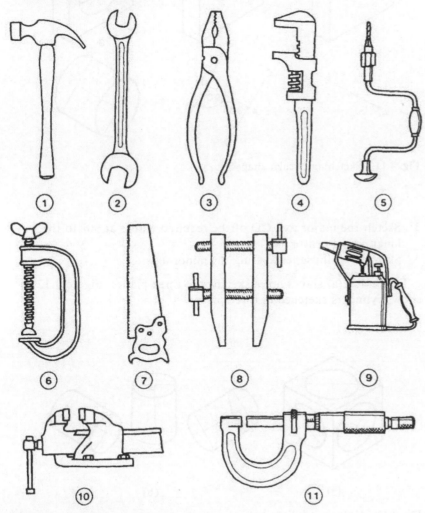

Fig. 4.7   Outline of engineering tools

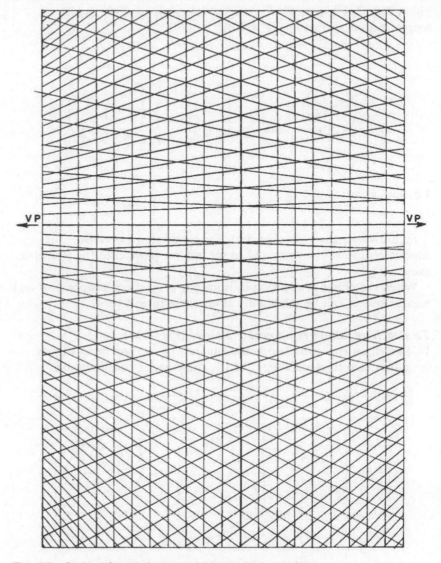

Fig. 4.8   Perspective-projection grid for guided sketching
To sketch an object as seen from above, place tracing paper or thin office paper over the grid below the horizon line.
To sketch the object as seen from below, use the top part of the grid.

4.3 Isometric projection

Isometric sketching starts with three basic axes equispaced as shown in Fig. 4.9(a). For practical reasons, the isometric axes are usually represented as shown in Fig. 4.9(b).

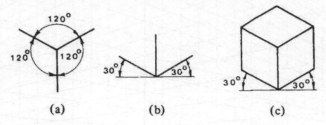

Fig. 4.9 Isometric axes

Figure 4.9(c) shows a cube drawn in isometric projection – the edges receding to the right and to the left are parallel to the isometric axes.

When sketching or drawing in isometric projection, proportions and measurements can be made only along these three axes.

To sketch an ellipse representing a circle (Fig. 4.10)

1 Sketch an enclosing 'isometric square', i.e. a rhombus, with its sides equal to the diameter of the circle under consideration.

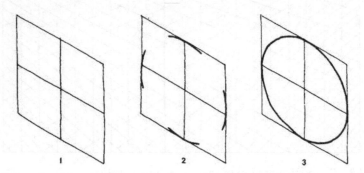

Fig. 4.10 Sketching a circle

2 Sketch bisecting lines and, at the intersection points, sketch short tangential arcs.
3 Finish the ellipse with a uniform bold line.

Alternative quick method of sketching ellipses (Fig. 4.11)

1 Sketch a faint construction line AB representing the longitudinal centre of a hole or cylinder.

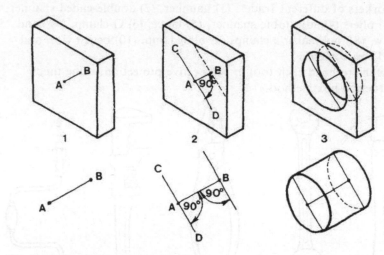

Fig. 4.11 Sketching circular shapes

2 Sketch the major axis CD of the required ellipse at 90° to the longitudinal centre line.
3 Sketch the ellipse, estimating the minor axis.

Figure 4.12(a) shows holes sketched in three planes. Figure 4.12(b) shows cylinders sketched in three planes.

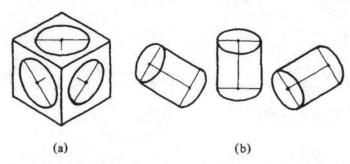

Fig. 4.12 Holes and cylinders sketched in three planes

To sketch cylindrical objects, first draw the complete construction ellipses, then draw tangential blending lines as shown in Fig. 4.13.

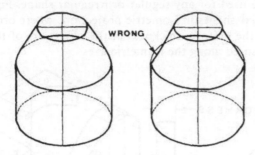

Fig. 4.13 Cylindrical objects

To sketch a nut (Fig. 4.14)

1 Sketch three vertical construction lines, placed at D and $D/2$ apart, where dimension D represents a nominal (major) thread diameter. Then sketch a hexagonal prism of height D, using the three vertical construction lines with all receding sides sloping at $45°$.

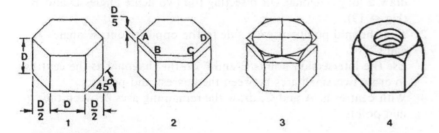

Fig. 4.14 Sketching a nut

2 Sketch lines parallel to the edges AB, BC and CD at a distance of $D/5$ from them, and sketch three arcs between these construction lines.
3 Sketch a horizontal tangential ellipse in the upper hexagon, representing the chamfer circle.
4 Sketch two blending arcs and include the threaded hole, which should be slightly smaller than D in diameter. Finish the sketch with uniform bold lines. The threaded hole would be represented by equispaced parallel sections of ellipses.

Fig. 4.15 Isometric grid for guided sketching

19

Drawing in isometric projection

A pictorial view of an object can be produced in isometric projection using drawing instruments. To overcome the effect of the receding lines appearing to be slightly larger than actual size, a reduced or isometric scale can be used, where receding lines are about 0.8 of the true lengths, Fig. 4.16. In practice, however, little use is made of this scale.

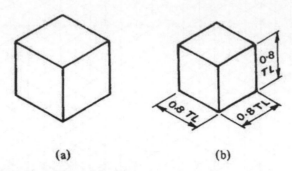

(a)   (b)

Fig. 4.16 (a) All actual true lengths
(b) Foreshortened receding lines

To draw an isometric circle using ordinates (Fig. 4.17)

1 Draw a circle as a plane figure, with centre lines inscribed in a square. Divide the circle by an even number of equidistant ordinates.
2 Draw the required 'isometric square' with all sides equal and add centre lines.
3 Transfer all ordinates from the plane-circle drawing to the 'isometric square' along the centre line AB, with corresponding ordinate measurements above and below AB.

4 Join the plotted points with a uniform bold line, preferably using a French curve to complete the ellipse.

This system of transferring ordinates from plane figures to isometric views may be used for any regular or irregular shape. Figure 4.18 shows irregular shapes in isometric projection, where ordinates representing the uniform thickness of the object are of the same length and are measured along the isometric axes.

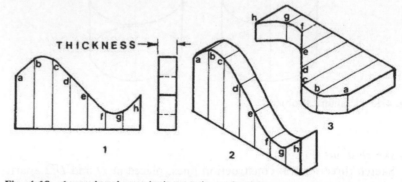

Fig. 4.18 Irregular shapes in isometric projection

To construct isometric circles using instruments (Figs 4.19 and 4.20)

1 Draw an 'isometric square' ABCD, each side being equal, and then draw a long diagonal DB bisecting the two acute angles D and B (Fig. 4.19).
2 Join the mid-points of each side to the opposite obtuse angles A and C.
3 Use the intersection points E and F on the diagonal as the centres to draw two small arcs between the nearest mid-points.
4 With centres at A and C, draw the remaining arcs between the mid-points.

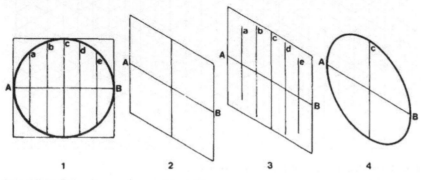

Fig. 4.17 Drawing an isometric circle

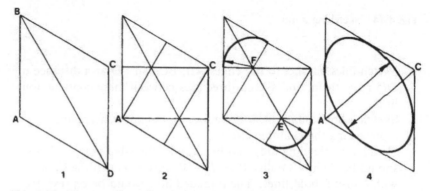

Fig. 4.19 Construction of a circle in isometric projection

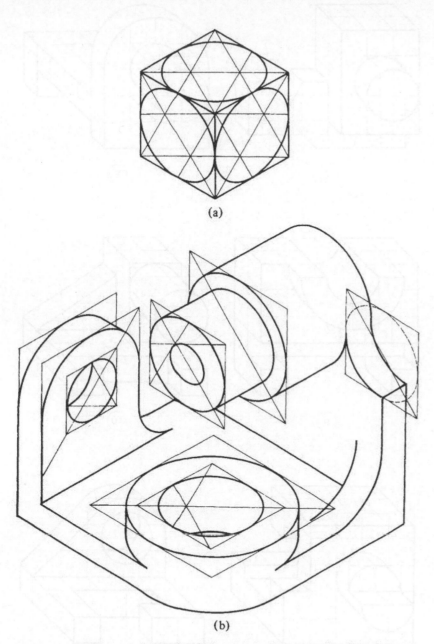

Fig. 4.20 Isometric circles in three planes: (a) all construction lines shown, (b) some construction lines omitted for clarity

4.4 Test questions

1 Redraw the components shown in Fig. 4.21 in isometric projection, remembering that measurements can be taken only along the basic axes. Each construction square represents a 10 mm measurement and should not be shown on your drawing.

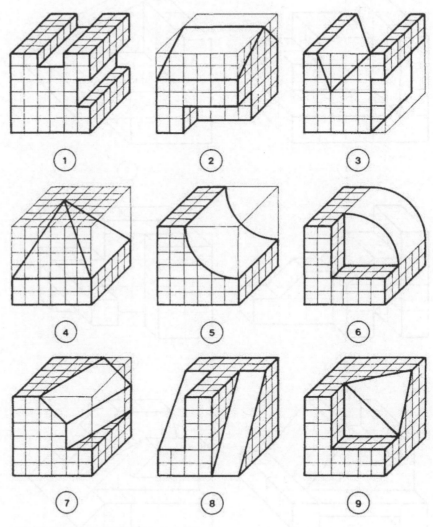

Fig. 4.21 Test question 1

2 Redraw in isometric projection the twelve objects shown in
Fig. 4.22, remembering that measurements can be taken only along
the basic axes. Each construction square represents a 10 mm
measurement.

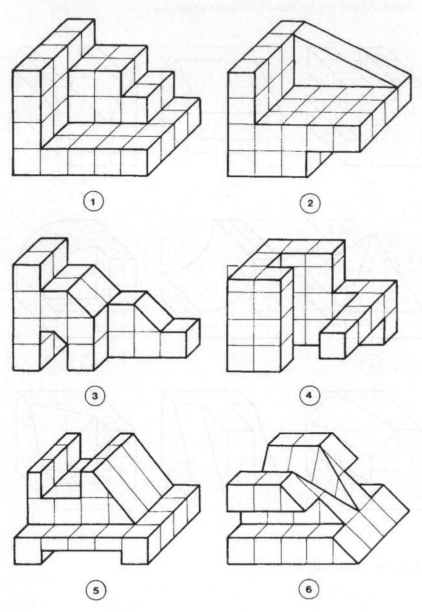

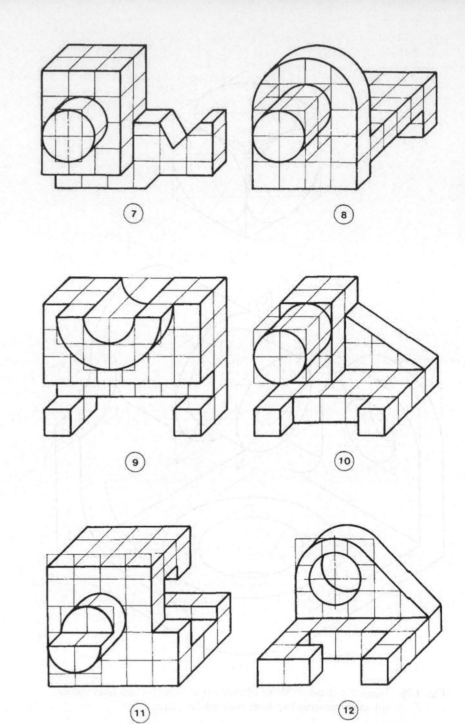

Fig. 4.22 Test question 2

4.5 Oblique projection

A pictorial view of an object can be produced in oblique projection, where the front face is sketched as a true shape without distortion.

Sketching in this projection is much easier than in isometric projection, since all the circles in the front face are sketched as plane figures instead of ellipses as in isometric projection.

Oblique sketching starts with two axes – one vertical and one horizontal – together with a third axis which is usually drawn at 45° to the horizontal and along which all measurements are reduced to half true length (TL), as shown in Fig. 4.23. This projection is sometimes called a 'cabinet' projection.

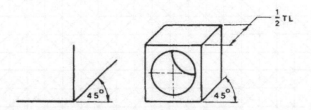

Fig. 4.23 Basic oblique axes and 'cabinet' projection

All proportions and measurements can be made only along these three axes.

A comparison of three pictorial projections is shown in Fig. 4.24.

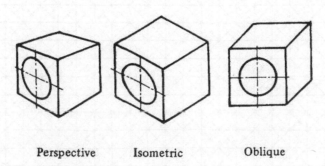

Perspective   Isometric   Oblique

Fig. 4.24 Pictorial projections

Drawing in oblique projection
The main advantage of oblique projection over isometric is that any complicated face – which may be curved, irregular, or have a number of holes – can be drawn as its true shape using drawing instruments.

The receding surfaces can be drawn at any angle but are usually drawn at 45° and are foreshortened to half true length.

All receding surfaces shown in Fig. 4.25 are half true length, but at different angles.

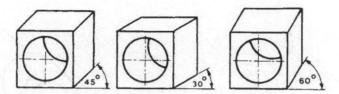

Fig. 4.25 Receding surfaces at different angles

In Fig. 4.26 the required semicircles are drawn with centres at A, B and C positioned along a line at 45° and at distances of half true lengths.

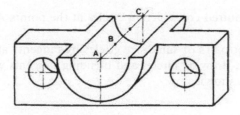

Fig. 4.26 Bracket in oblique projection

Figure 4.27 indicates the advantages of oblique projection when drawing a complicated cylindrical component.

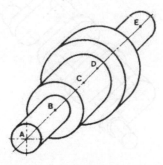

Fig. 4.27 Cylindrical component in oblique projection

To draw cylindrical shapes in oblique projection (Fig. 4.28)

1　Draw the axis of the cylindrical shape at 45° to the horizontal and locate the points A, B, C, and D representing all normal (at 90°) surfaces along the axis at distances of half true length.

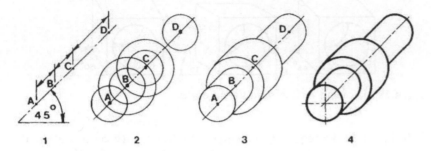

Fig. 4.28　Drawing cylindrical shapes

2　Draw the required construction circles at the points A, B, C, and D.

3　Join adjacent pairs of same-size circles by tangents at 45°.

4　To complete, draw all circles and blending tangents with uniform bold lines as shown.

Figure 4.29 shows how complicated cylindrical shapes can be represented in oblique projection.

Fig. 4.29　Crankshaft in oblique projection

Fig. 4.30　Oblique grid for guided sketching

24

4.6 Test questions

1 Redraw in oblique projection the objects shown in Fig. 4.31, remembering that measurements can be taken only along the basic axes. Each construction square represents a 10 mm measurement.

2 Redraw in oblique projection the objects shown in Figure 4.32, remembering that measurements can be taken only along the basic axes. Each construction square represents a 10 mm measurement.

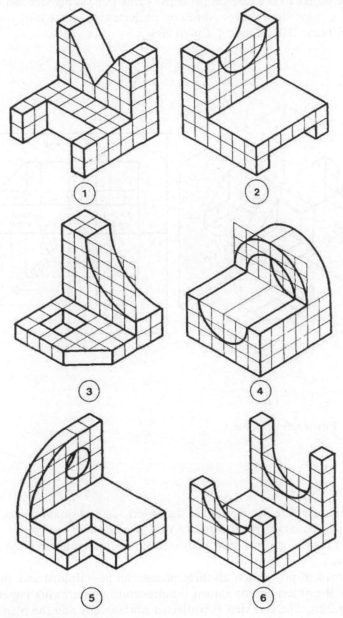

Fig. 4.31 Test question 1

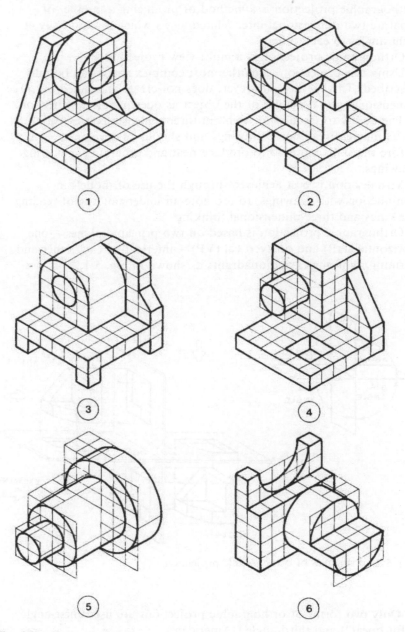

Fig. 4.32 Test question 2

5 Orthographic projection

Orthographic projection is a method of producing a number of separate two-dimensional inter-related views which are mutually at right angles to each other.

Orthographic projection is a multi-view projection.

Using this projection, even the most complex shape can be fully described. This method, however, does not create an immediate three-dimensional visual picture of the object as does pictorial projection.

The ability to visualise or think in three dimensions is essential to the competent reading of drawings and should be developed even before the skills required to produce neat and accurate engineering drawings.

Visualisation is best achieved through the use of models in conjunction with drawings, to promote an understanding of reading drawings and three-dimensional thinking.

Orthographic projection is based on two principal planes – one horizontal (HP) and one vertical (VP) – intersecting each other and forming right angles and quadrants as shown in Fig. 5.1.

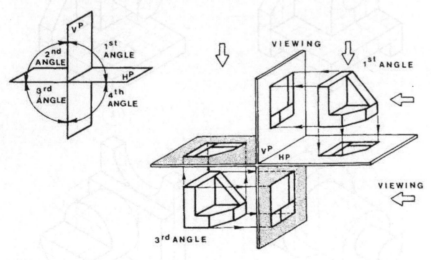

Fig. 5.1 Principles of orthographic projection

Only two forms of orthographic projections are used: first-angle ('European') and third-angle ('American').

5.1 First-angle projection

In first-angle projection, an object is positioned in the space of the first-angle quadrant between two planes Fig. 5.2(a). A view of the object is projected by drawing parallel projecting lines, or projectors, from the object to the vertical principal plane (VP). This view on VP is called a front view. A view similarly projected on to the horizontal principal plane (HP) is called a plan view.

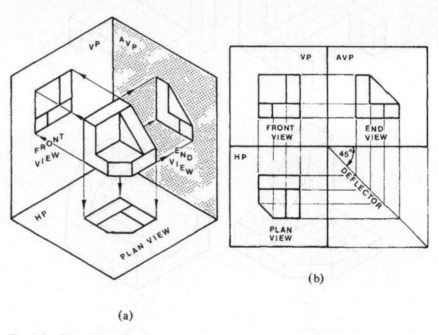

(a)

(b)

Fig. 5.2 First-angle projection

For the complete description of the object, an additional plane, called the auxiliary vertical plane (AVP), is used at 90° to the principal planes, and the view projected on to that plane is called an end view.

By means of projectors, all three planes can be unfolded and three views of the object can be shown simultaneously on drawing paper as in Fig. 5.2(b). The end view is projected horizontally and the plan view vertically from the front view.

In first-angle projection, the object always comes between the eye of the observer and the projection plane or view, as shown in Fig. 5.3.

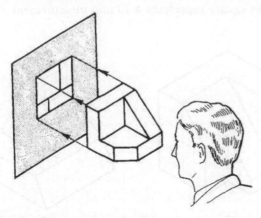

Fig. 5.3 Principle of first-angle projection

The symbol used to indicate first-angle projection is derived from views of a circular taper as shown in Fig. 5.4. The symbol shows a front view and a left end view of the circular taper in first-angle projection. It should be drawn in proportion, in terms of taper diameter d as shown.

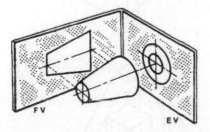

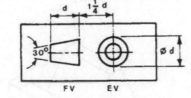

Fig. 5.4 First-angle projection symbol

Sometimes it is necessary to show six views of an object, as in Fig. 5.5. To show hidden detail, a thin line with short dashes is used.

As a rule, the minimum number of views should be used, especially to represent simple objects. The views should be selected so that they clearly indicate all the required detail.

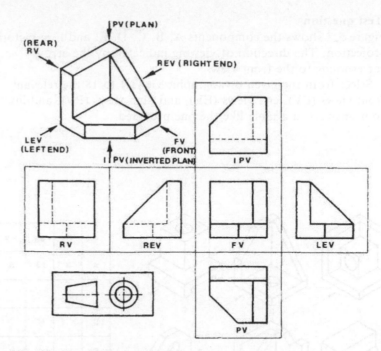

Fig. 5.5 Six views in first-angle projection

In common practice only three views are used: (a) a front view, (b) an end view, and (c) a plan view. These three views are sufficient for a complete description of an object, as shown in Fig. 5.6.

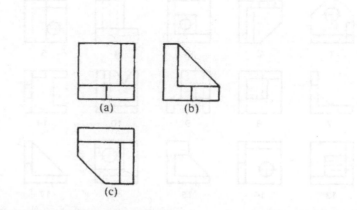

Fig. 5.6 Three views in first-angle projection

27

5.2 Test questions

1 Figure 5.7 shows the components A, B, C, D, E, and F in pictorial projection. The direction of viewing indicated by the arrow corresponds to the front view.

  Select from the given orthographic views 1 to 18 the relevant front views (FV), end views (EV), and plan views (PV), and insert your answers in a table like the one provided.

2 Draw or sketch, full size, in first-angle projection the components shown in Fig. 5.8. Select the views to show most of the features as visible outlines, and include hidden detail where necessary. Each construction square represents a 10 mm measurement.

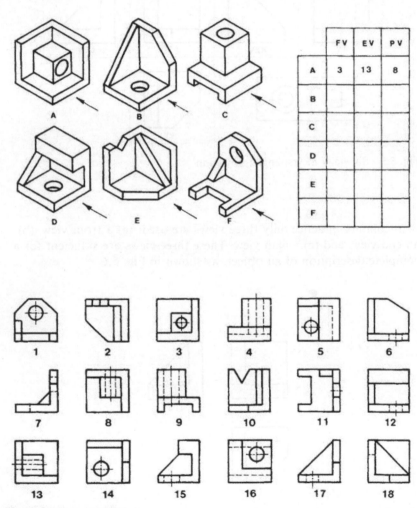

	FV	EV	PV
A	3	13	8
B			
C			
D			
E			
F			

Fig. 5.7 Test question 1

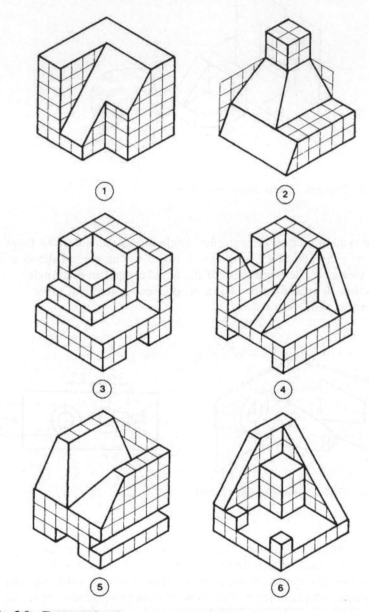

Fig. 5.8 Test question 2

28

3 Complete the third missing view for each object shown in first-angle projection in Fig. 5.9. Squared or tracing paper may be used.

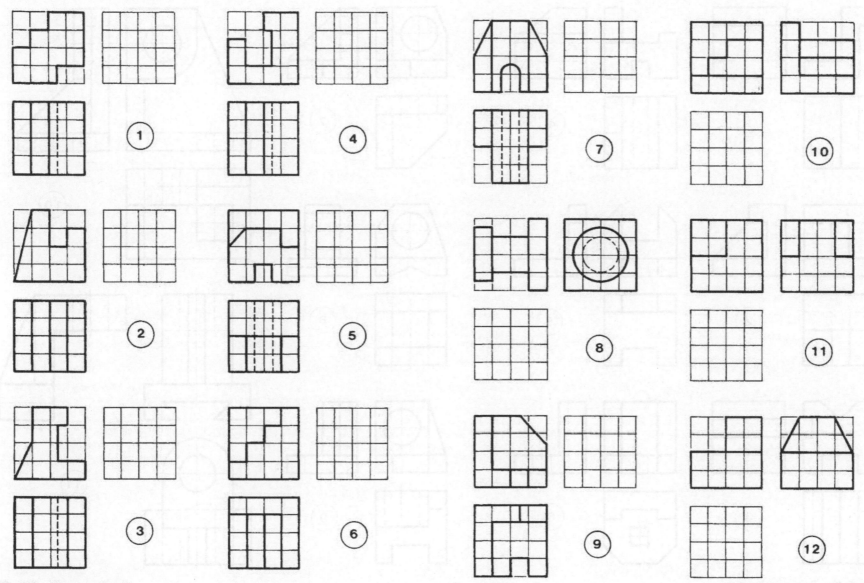

Fig. 5.9 Test question 3

4 The components shown in Fig. 5.10 are drawn in half-size in first-angle projection. Sketch or redraw each component
(a) in isometric projection,

(b) in oblique projection, where the position of component 11 may be rearranged.
Isometric and oblique grids may be used (pages 19 and 24)

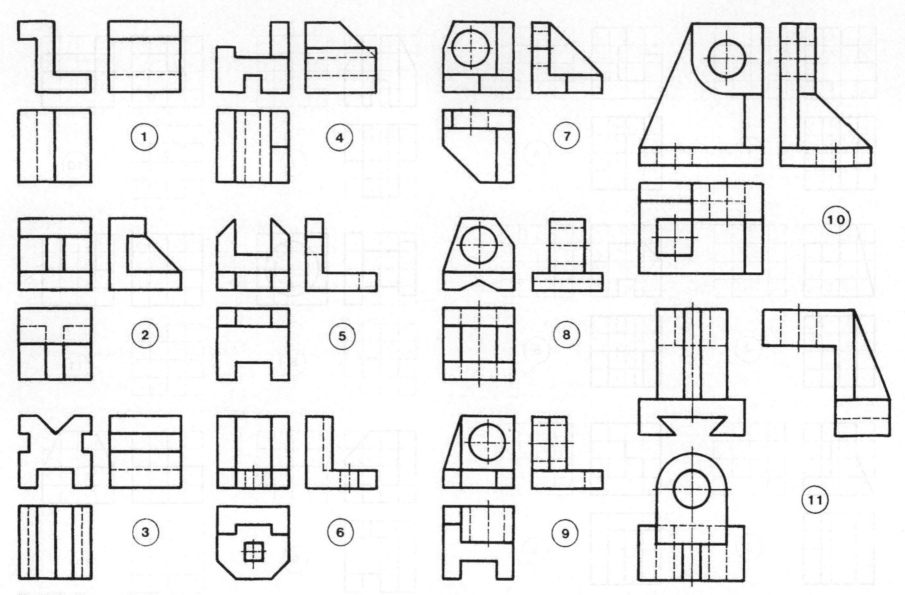

Fig. 5.10 Test question 4

5.3 Third-angle projection

In third-angle projection, an object is positioned in the space of the third-angle quadrant between two principal planes. The planes are imagined to be transparent, and the projected views of the object are viewed through the planes as shown in Fig. 5.12(a).

By means of projectors, all three planes of the 'glass box' can be unfolded and three views of the object can be shown simultaneously on drawing paper as in Fig. 5.12(b).

In both first- and third-angle projection the views are identical, but the positioning of each is different.

In third-angle projection, the 'transparent' projection plane or view always comes between the eye of the observer and the object, as shown in Fig. 5.13.

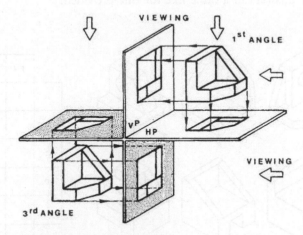

Fig. 5.11 Principle of orthographic projection

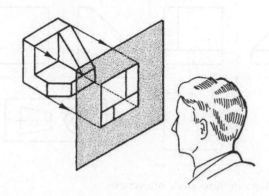

Fig. 5.13 Principle of third-angle projection

The symbol used to indicate third-angle projection is derived as for first-angle projection, but the views are positioned differently, as shown in Fig. 5.14. The symbol shows a left end view and a front view of the circular taper in third-angle projection. See p. 27, Fig. 5.4 for recommended proportions.

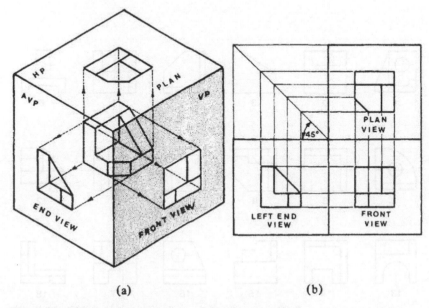

Fig. 5.12 Third-angle projection – 'glass-box' method

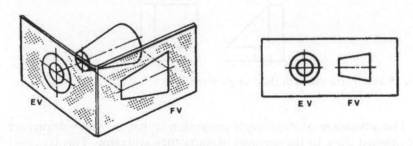

Fig. 5.14 Third-angle projection symbol

31

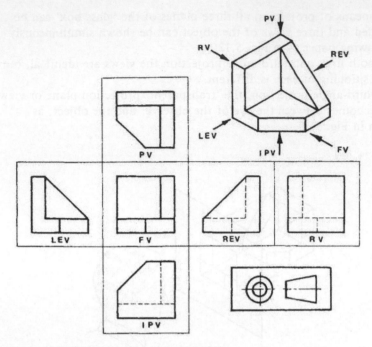

Fig. 5.15 Six views in third-angle projection

In common practice, only three of the six possible views (Fig. 5.15) are used for a complete description of an object, as in Fig. 5.16. Either of the two end views may be chosen.

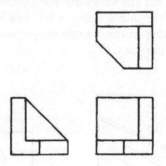

Fig. 5.16 Three views in third-angle projection

The advantage of third-angle projection is that the views drawn are positioned close to the surfaces or parts they represent. This is specially relevant with larger drawings.

5.4 Test questions

1 Figure 5.17 shows components A, B, C, D, E, and F in pictorial projection. The direction of viewing indicated by the arrows corresponds to the front views.

Select from the given orthographic views 1 to 18 the relevant front views (FV), end views (EV), and plan views (PV), and insert your answers in a table like the one provided.

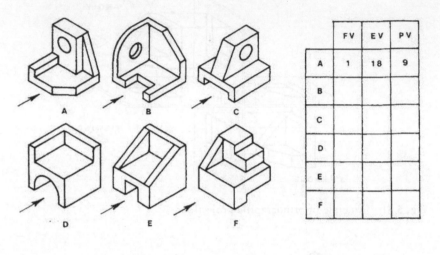

	FV	EV	PV
A	1	18	9
B			
C			
D			
E			
F			

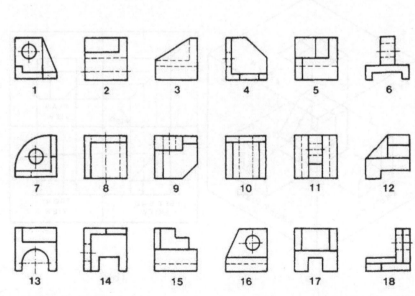

Fig. 5.17 Test question 1

32

2 (a) Draw or sketch in third-angle projection the six views of the component shown in Fig. 5.18(a). Name each view in full.
   (b) Draw or sketch in third-angle projection the six views of the component shown in Fig. 5.18(b).
   (c) Draw or sketch in first-angle projection the six views of the component shown in Fig. 5.18(c).
   Each construction square represents a 10 mm measurement.

3 Draw or sketch in third-angle projection the components shown in Fig. 5.19. Select views to show most of the features as visible outlines. Each construction square represents a 10 mm measurement. Include hidden detail.

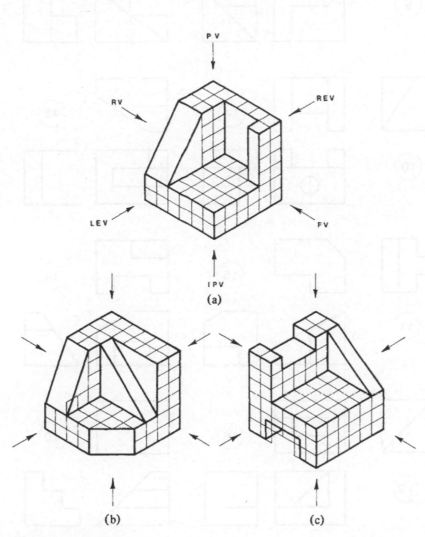

Fig. 5.18 Test question 2

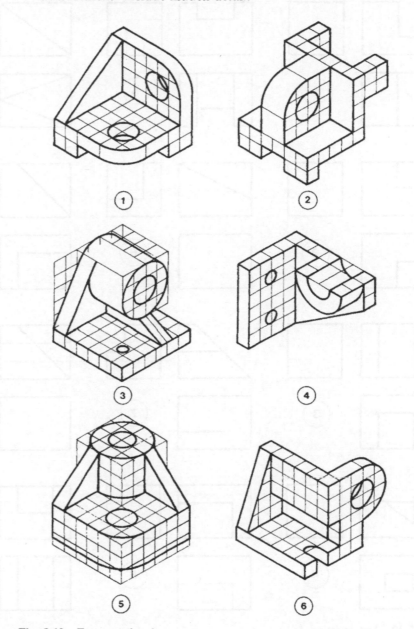

Fig. 5.19 Test question 3

4 In Fig. 5.20, incomplete drawings 1 to 12 are in first-angle projection and 13 to 20 in third-angle. Complete each drawing by adding the missing lines. Tracing paper may be used.

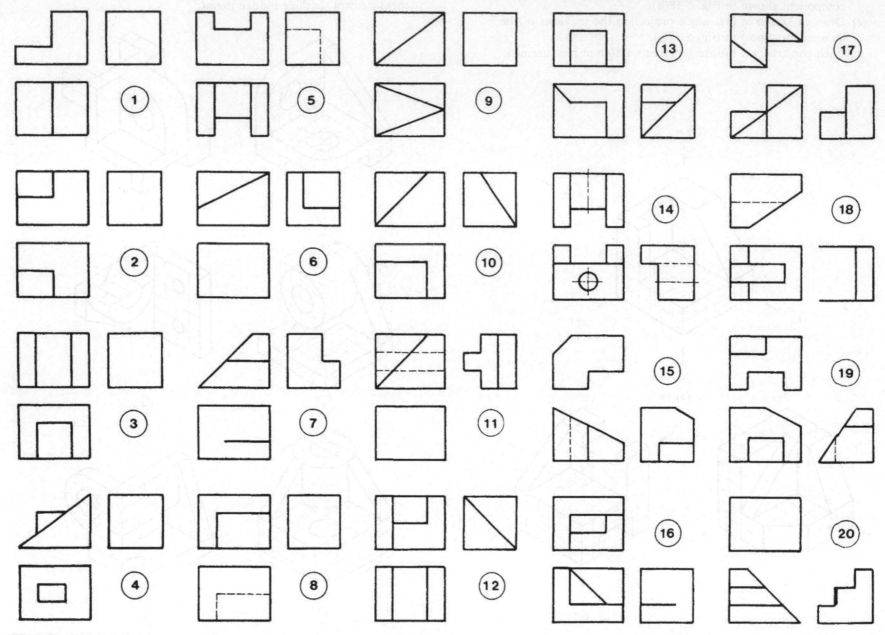

Fig. 5.20 Test question 4

5 Complete the third view for each object shown in third-angle projection in Fig. 5.21. Squared or tracing paper may be used.

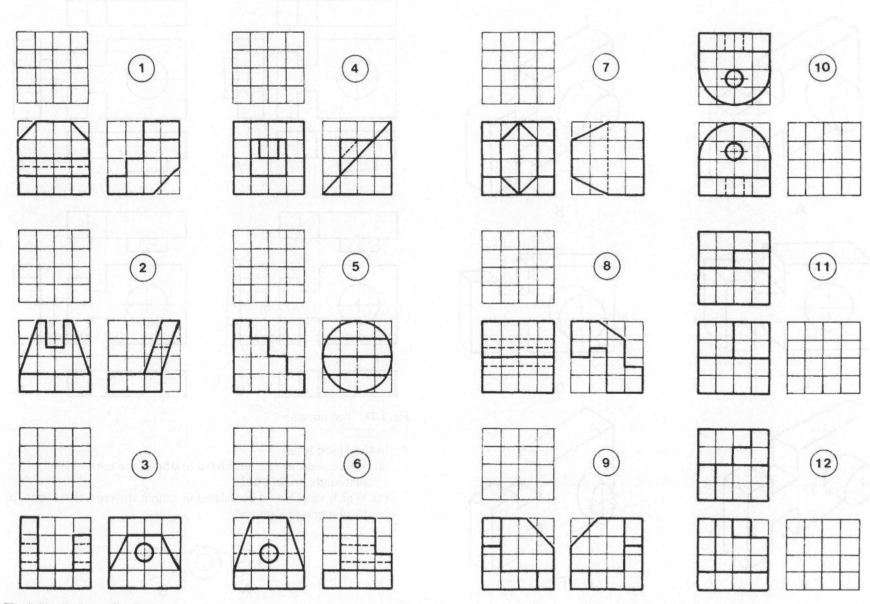

Fig. 5.21 Test question 5

6 Figure 5.22 shows a component in three different pictorial projections.
   (a) Which is the correct perspective projection?
   (b) Which is the correct isometric projection?
   (c) Which is the correct oblique projection?

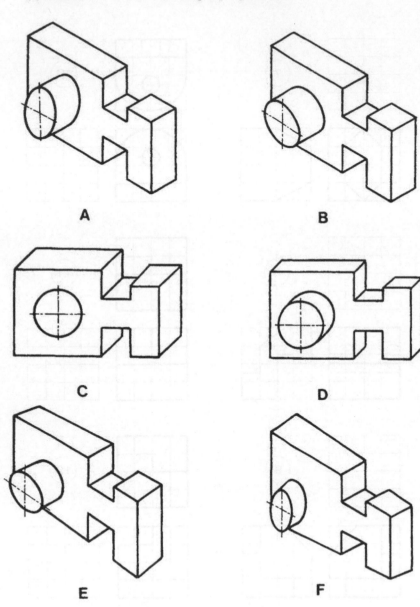

A   B

C   D

E   F

Fig. 5.22 Test question 6

7 Look at the drawings of a component shown in Fig. 5.23.
   (a) Which is in correct first-angle projection?
   (b) Which is in correct third-angle projection?

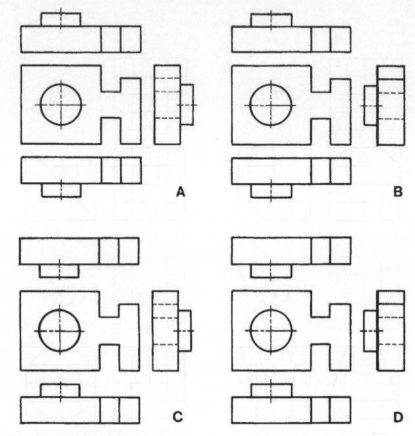

A   B

C   D

Fig. 5.23 Test question 7

8 In the figure below,
   (a) Which view has to be deleted to obtain the correct symbol for first-angle projection?
   (b) Which view has to be deleted to obtain the correct symbol for third-angle projection?

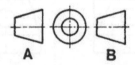

A   B

Fig. 5.24 Test question 8

9 Select the correct end view – A, B, C, or D – for each object shown in Fig. 5.25.

10 Sketch in first-angle projection objects 1 to 6 shown in Fig. 4.22 on page 22.
11 Sketch in third-angle projection objects 7 to 12 shown in Fig. 4.22 on page 22.
12 Draw in first-angle projection the objects shown in Fig. 4.31 on page 25.
13 Draw in third-angle projection the objects shown in Fig. 4.32 on page 25.

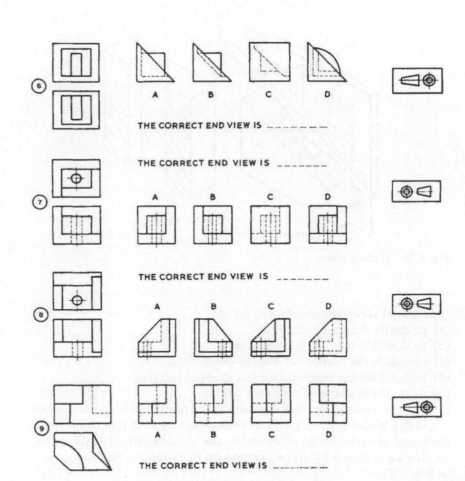

THE CORRECT END VIEW IS _____

Fig. 5.25   Test question 9

5.5 Sectional views

Quite often an outside view of an object does not adequately describe it, as no internal features are shown.

In order to show the internal features without excessive use of hidden-detail lines, the object is imagined to be cut along a plane called a *cutting plane*. The cut portion nearer to the observer is removed and the remaining part is shown as a sectional view.

The surfaces in section can be imagined to be cut along the cutting plane with an imaginary tool and imaginary cutting marks are represented by thin equidistant hatching lines as shown in Fig. 5.26. Sometimes hatching may be omitted, if the clarity of drawing is not reduced by doing so.

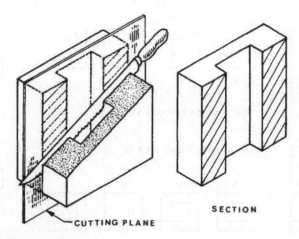

Fig. 5.26   Cutting plane

Sectional views are usually produced
(a) to clarify details of the object,
(b) to illustrate internal features clearly,
(c) to reduce the number of hidden-detail lines,
(d) to facilitate the dimensioning of internal features,
(e) to show the shape of the cross-section,
(f) to show clearly the relative positions of parts forming an assembly.

Cutting planes are represented on drawings by long thin chain lines thickened at each change of direction and at both ends. The direction of viewing is shown by arrows resting on thick lines at both ends, as in Fig. 5.27.

Cutting planes should be designated by capital letters.

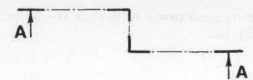

Fig. 5.27   Cutting plane AA

The surfaces shown in section are usually hatched at 45° or at some well defined angle which avoids clashing with visible outlines, as in Fig. 5.28. Spacing between hatching lines should be equidistant and not less than 4 mm.

Fig. 5.28   Hatching

Adjacent components should be hatched in opposite directions. Hatching lines for additional adjacent parts can be offset or, alternatively, spacing between the lines may be increased or reduced, as shown for the internal part in Fig. 5.29(a) and (b).

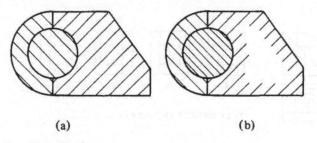

(a)   (b)

Fig. 5.29   Hatching of adjacent parts

Spacing between the hatching lines should be chosen in proportion to the size of the hatched section. In the case of large areas in section, the hatching may be limited to a zone following the contour of the hatched area, as shown in Fig. 5.29(b).

When sections are shown side by side in parallel planes, as in Fig. 5.30, hatching lines should not indicate any change in direction of the cutting plane.

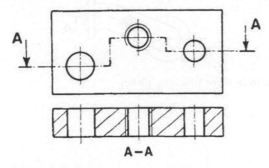

Fig. 5.30 Parallel cutting planes

Thin sections should be shown as single thick lines, leaving space between adjacent parts for clarity, as shown in Fig. 5.31(a).

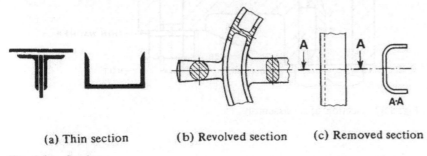

(a) Thin section   (b) Revolved section   (c) Removed section

Fig. 5.31 Sections

Cross-sections may be revolved in place, as in Fig. 5.31(b). The outline is shown in continuous thin lines, and further identification is not necessary.

A cross-section may be removed as shown in Fig. 5.31(c) — further identification is then necessary.

A symmetrical component in Fig. 5.32(a) is shown in half section. The centre line, which separates the outside view from the sectional view *always* takes precedence over any other form of line. (see p. 4).

A local (part) section may be drawn to avoid the need for a separate sectional view. The local break is shown by a continuous thin irregular freehand line, as shown in Fig. 5.32(b).

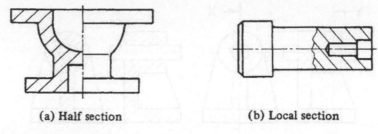

(a) Half section   (b) Local section

Fig. 5.32 More sections

There are several exceptions to the general rules of sectioning, as follows.

For clarity, the following features are not shown in section when cut longitudinally:
(a) ribs and webs,
(b) shafts, rods and spindles,
(c) bolts, nuts, and thin washers,
(d) rivets, dowels, pins and cotters.
(e) spokes of wheels and similar parts.

If a shaft and web lie along the cutting plane, they are not sectioned, Fig. 5.33(a). If a shaft and web lie across the cutting plane, as in Fig. 5.33(b), then they are sectioned.

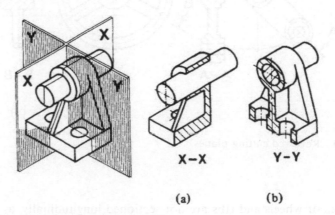

(a)   (b)

Fig. 5.33 Cutting planes XX and YY

The same sections on XX and YY are shown in orthographic projection in Fig. 5.34.

39

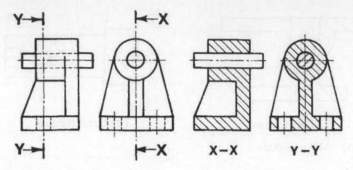

Fig. 5.34 Sections X–X and Y–Y

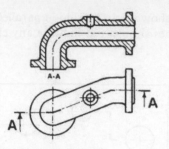

Fig. 5.36 Section in three touching planes

In Fig. 5.35(a), the cutting plane is revolved into the vertical position and is then projected to the sectional view.

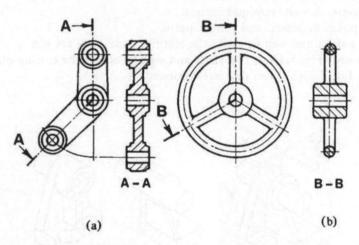

(a)

(b)

Fig. 5.35 Revolved cutting planes

Spokes of wheels and ribs are not sectioned longitudinally, as shown in Fig. 5.35(b). The cutting plane is revolved into the vertical position and is then projected to the sectional view.

Figure 5.36 shows a section in three touching adjoining planes.

Figure 5.37 shows a section through an assembly which includes the features that are not usually sectioned.

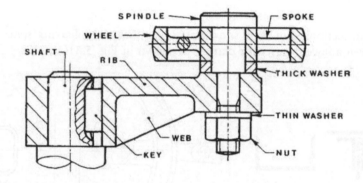

Fig. 5.37 Section of an assembly

When webs or ribs are cut along their length by a cutting plane, they are not sectioned, in order to avoid a false appearance of solidity. The webs and ribs are usually thin in comparison to the overall thickness of the main body.

If a cutting plane cuts across the webs or ribs, then they are shown in section in the usual way.

Nuts and bolts, thin washers, studs, screws, rivets, keys, pins, shafts, spindles, and spokes of wheels are more easily recognisable by their external features, so they are not shown in section if cut longitudinally (Fig. 5.37).

40

5.6 Test questions

1 The components shown in Fig. 5.38 are drawn in first- or third-angle projection. Sketch or redraw the two given views of each component and complete the third as a sectional view. Tracing paper may be used.

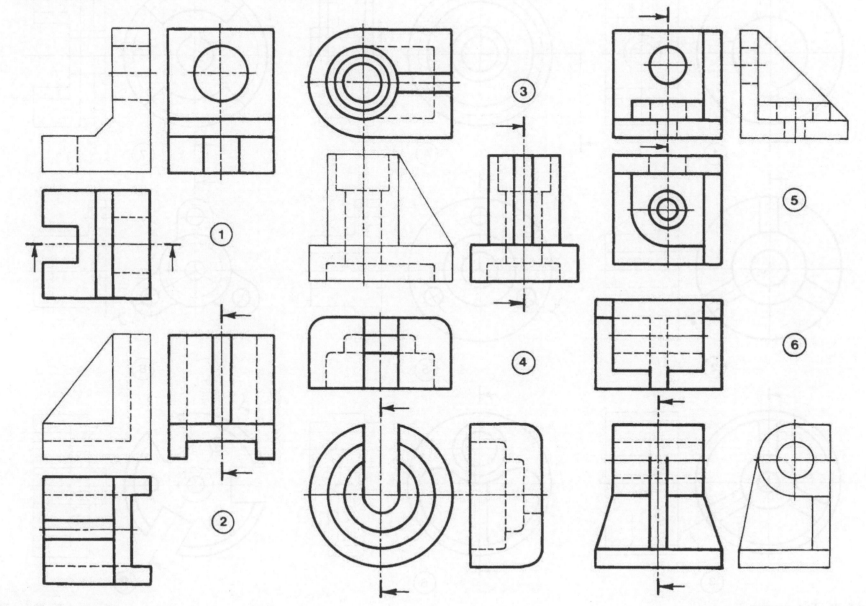

Fig. 5.38 Test question 1

2 For the components shown in Fig. 5.39 complete all the end views in section as indicated. Tracing paper may be used.

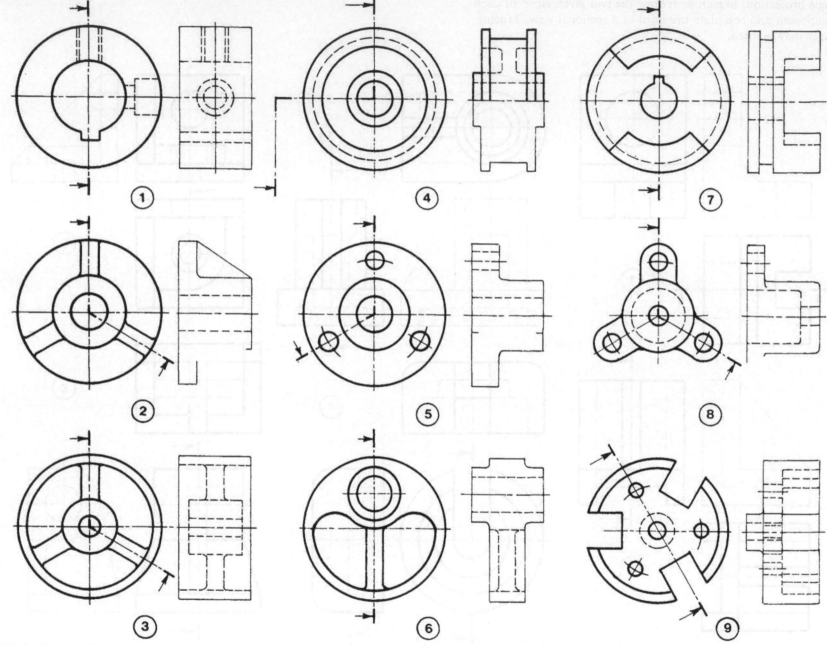

Fig. 5.39 Test question 2

3 Draw or sketch twice full size (2:1) in first-angle projection the following views of the bracket shown in Fig. 5.40:
(a) a sectional front view on AA,
(b) a sectional plan view on BB,
(c) an end view.
Ensure the correct positioning of all views.
Each construction square represents a 5 mm measurement.

4 Draw or sketch half full size (1:2) in third-angle projection the following views of the bracket shown in Fig. 5.41:
(a) a sectional front view on AA,
(b) an end view,
(c) a plan.
Ensure the correct positioning of all views
Each construction square represents a 20 mm measurement.

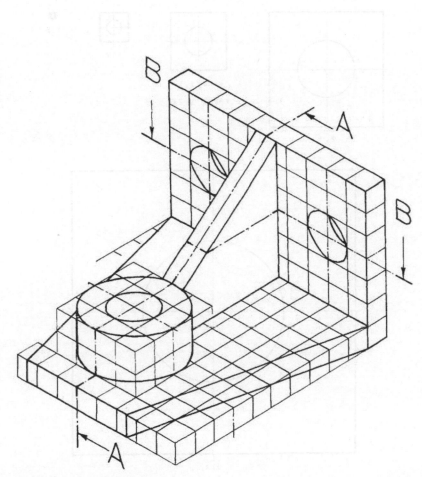

Fig. 5.40 Test question 3

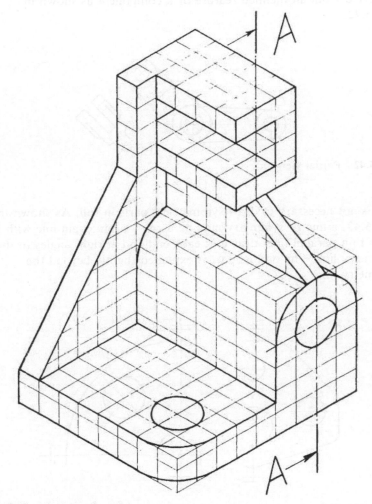

Fig. 5.41 Test question 4

5.7 Views on drawings

To ensure clear reading, due regard should be paid to the spacing of views on drawings – views and sections should not be overcrowded.

The number of views should be the minimum necessary to ensure that the drawing can be fully understood.

Views should be chosen so that as little information as possible has to be shown as hidden detail and to provide clearly and precisely the maximum possible information. In general, hidden detail should be used only where it is essential, and it should not be used for dimensioning purposes.

It is sometimes necessary to draw a partial view, which may be projected from an inclined feature of a component as shown in Fig. 5.42.

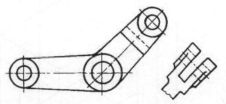

Fig. 5.42   Partial view

It is not necessary to draw symmetrical parts in full. As shown in Fig. 5.43, a line of symmetry may be used – a thin chain line with two short thin parallel lines drawn at each end and at right angles to the symmetry line. The outline part is extended slightly beyond the symmetry line.

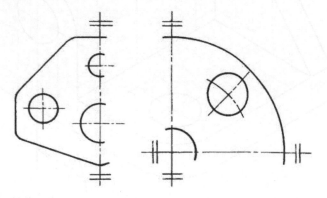

Fig. 5.43   Indicating symmetry of parts

Scale

Very large components are usually drawn to a reduced scale. To ensure clarity and precision, very small components are drawn larger than full size.

The drawing should be drawn in selected proportion to a uniform scale. This scale should be stated on the drawing as a ratio.

Scale multipliers and divisors of 2, 5, and 10 are recommended:

full size – 1:1

smaller than full size – 1:2, 1:5, 1:10, 1:20, 1:50, 1:100

larger than full size – 2:1, 5:1, 10:1, 20:1, 50:1, 100:1.

Some of the scales used are illustrated below.

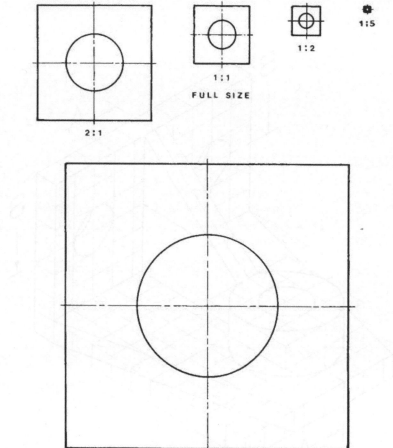

Fig. 5.44   Different scale ratios

5.8 Reading an engineering drawing

To read a drawing is to obtain a clear mental picture of what the person who prepared the drawing wishes to convey. Every engineer must be able to read and understand drawings.

In orthographic projection, at least two views are required for a full description of an object. Figure 5.45 shows that two given views do not necessarily describe an object completely, as several end views are possible.

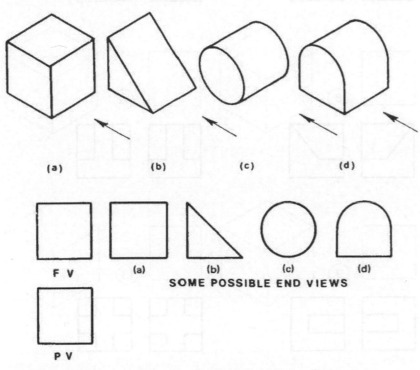

Fig. 5.45 Some possible end views for the given front and plan views

It is essential to acquire the ability to read drawings and to *visualise* the objects they represent. A drawing must be read patiently by referring systematically back and forth from one view to another. At the same time, the reader must imagine a three-dimensional object and not a two-dimensional flat projection.

A pictorial sketch usually helps to clarify the shape of a part that is difficult to visualise.

5.9 Test questions

1 Figure 5.46 shows two views in orthographic projection of each of six objects. Copy the given views using tracing paper, then complete the third views given and sketch each object in a pictorial projection in the space indicated.

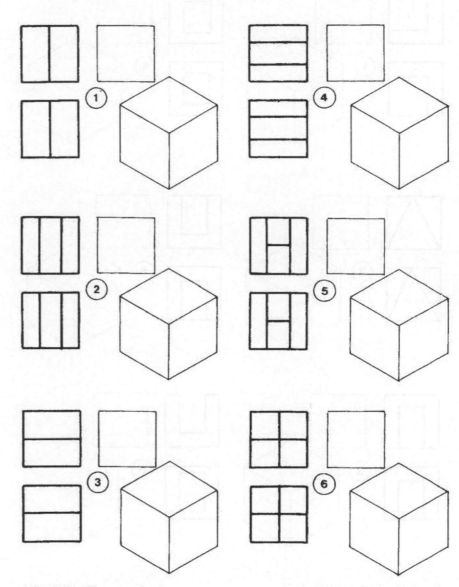

Fig. 5.46 Test question 1

45

2 For each of the objects 1 to 14 shown in orthographic projection in Fig. 5.47, copy the given views using tracing paper, then complete the unfinished third view and sketch the object in a pictorial projection.

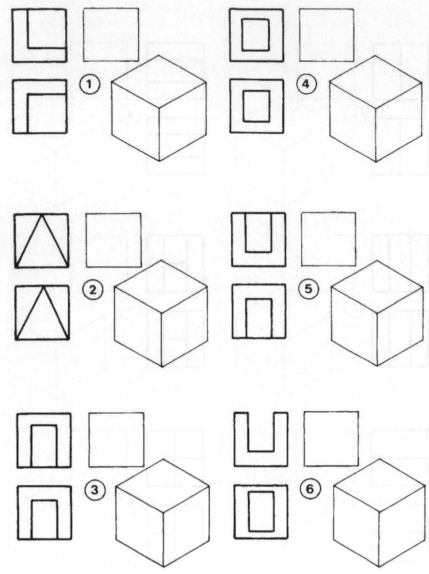

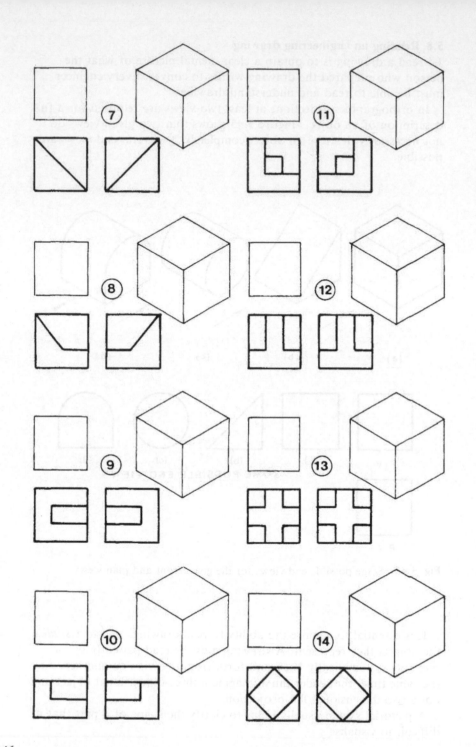

Fig. 5.47 Test question 2

3 Sketch in isometric or perspective projection each object shown in Fig. 5.48 and label all surfaces. Some surfaces may have more than one letter. The point W is to be in the lowest position.

Tracing paper may be used.

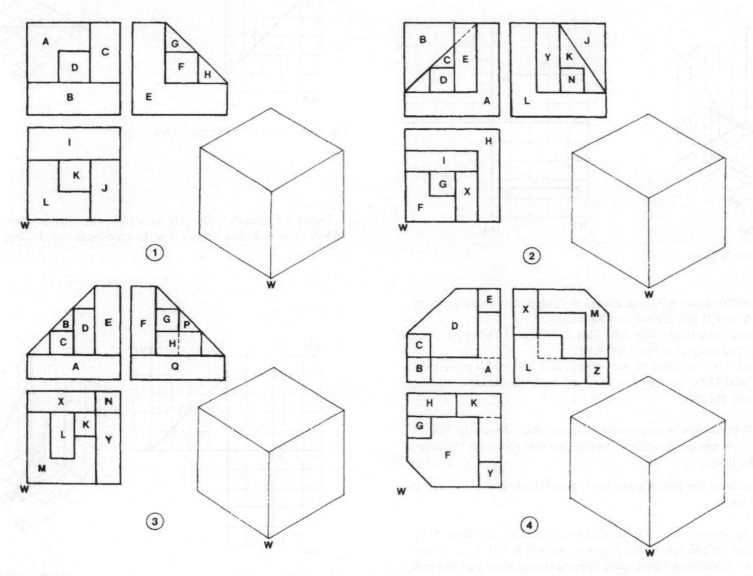

Fig. 5.48 Test question 3

5.10 Points, lines, and plane surfaces in space

Figure 5.49 (a) shows two points A and B suspended in the space between three principal planes drawn in first-angle projection.

The views of each point are projected to the principal planes and are indicated by the lower-case letters a and b, as if they were shadows of the points A and B.

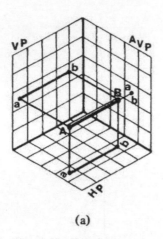

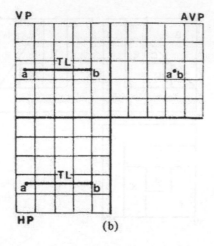

(a)

Fig. 5.49   Line in space

Figure 5.49(b) shows all three planes unfolded, and three views of the points A and B are shown simultaneously.

Now consider a straight line AB. This is projected as the lines ab on VP and HP but only as a point on AVP.

As the line AB is parallel to the vertical and horizontal principal planes (VP and HP), the views in those planes represent the true length (TL) of the line.

The rule   When a line is suspended in space, the view of the line projected on to any plane parallel to the line will represent the true length of the line.

Note   For clarity the line AB has been given thickness, as though it were a thin rod.

Figure 5.50 shows a line AB parallel to the horizontal plane (HP). The plan view on the horizontal plane is the true length (TL) of the line. The two remaining views show foreshortened lines and not true lengths.

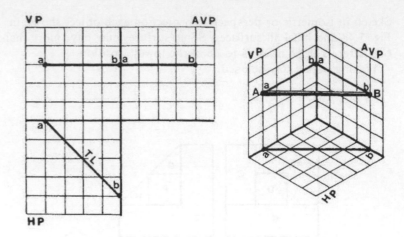

Fig. 5.50   Horizontal line suspended in space

Figure 5.51 shows a line AB parallel to the auxiliary vertical plane (AVP). The end view on AVP is the true length of the line.

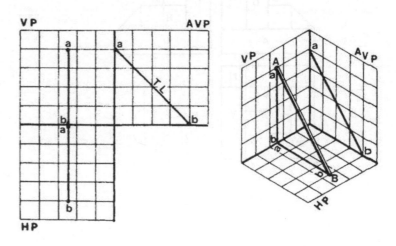

Fig. 5.51   Sloping line suspended in space

Figure 5.52 shows a line AB which is not parallel to any of the principal planes, therefore none of the views on these planes represents the true length of the line.

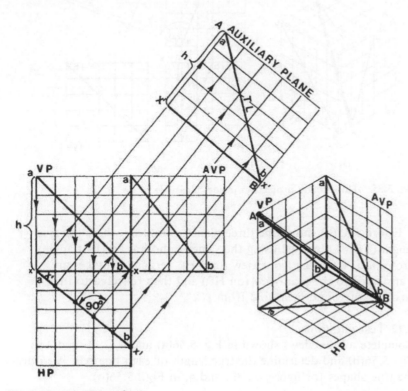

Fig. 5.52 Auxiliary plane

If the true length is to be shown, a new plane parallel to the line AB has to be introduced. This plane is called an *auxiliary plane*.

To draw the true length

1 Draw an auxiliary plane parallel to the line ab in the plan view on (HP), projected at right angles to that line.
2 Transfer all vertical distances h measured between the reference line xx and the line ab in the front view on (VP) to the auxiliary plane by reflecting at 90° from the line ab in the plan view (on HP) and measuring from x'x'.
3 The auxiliary view on AVP will be the true length of the line AB.

Figure 5.53 shows two lines AB and AC which are not parallel to any of the principal planes.

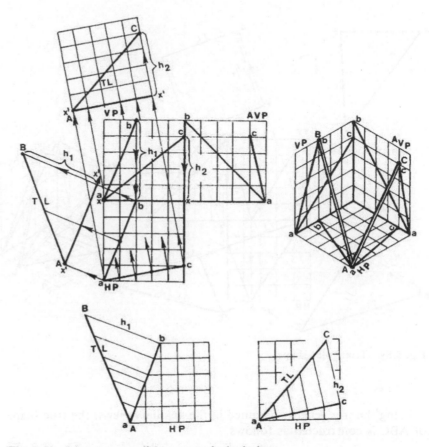

Fig. 5.53 Lines not parallel to any principal plane

To find the true lengths of the lines AB and AC, the required distances are transferred from the front view (on VP) to the plan view (on HP) and then reflected to the respective auxiliary views.

It is not necessary to draw a complete auxiliary plane in each case – it may be omitted as shown for the line AB, where only the distances from the reference line x'x' are used.

Figure 5.54 shows a triangular shape ABC. The true lengths are obtained by transferring all required vertical distances from the front view (on VP) to the auxiliary views.

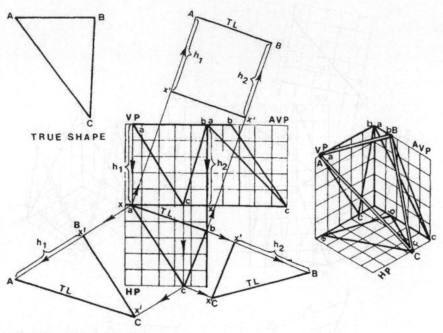

Fig. 5.54 Triangular shape

Using the true lengths obtained in the auxiliary views, the true shape of ABC is constructed as follows.

1 Draw the true length AB.
2 From centre A with true length AC as radius, strike an arc.
3 From centre B with true length BC as radius, strike an arc.
4 The intersection of these arcs is the required point C.
5 Join AC and BC.

Figure 5.55(a) shows the line AB drawn in third-angle projection inside a 'glass box'. All the projected views of the line AB are viewed through the 'transparent' principal planes.

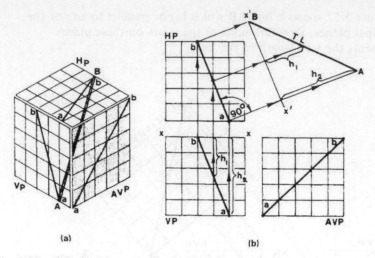

Fig. 5.55 Third-angle projection inside 'glass-box'

Figure 5.55(b) shows the three planes unfolded, where the true length (TL) is obtained using the same method as for first-angle projection: from the front view (on VP), vertical distances are transferred to the plan view (on HP) and then reflected to the auxiliary view and measured from x'x'.

5.12 Test questions

Complete all the views shown in Fig. 5.56(a) and 1, 2, 3, and 4 in Fig. 5.56(b) and determine the true length of each line AB. Also draw the true shapes for figures 2, 3b, and 4, in Fig., 5.56(b).

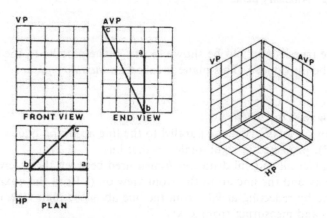

Fig. 5.56(a) Views for test questions

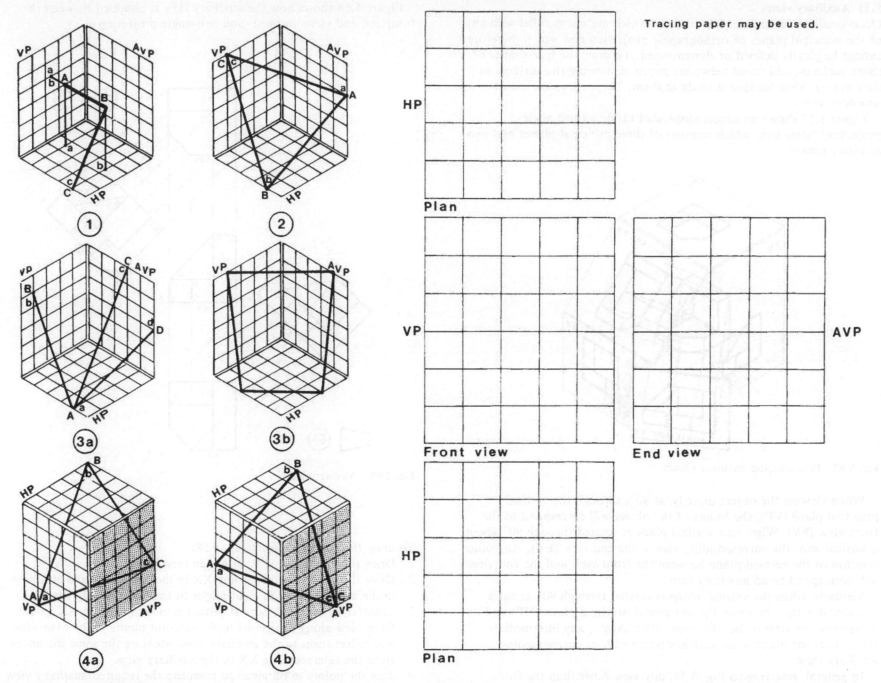

Tracing paper may be used.

Fig. 5.56(b) Views for test questions

51

5.11 Auxiliary views

Occasionally a component has surfaces which are not parallel with any of the principal planes of orthographic projection and which therefore cannot be clearly defined or dimensioned. To draw the true shapes of those surfaces, additional views are required showing the surfaces as they appear when looking directly at them. These views are called *auxiliary views*.

Figure 5.57 shows an object suspended inside a third-angle-projection 'glass box' which consists of three principal planes and two auxiliary planes.

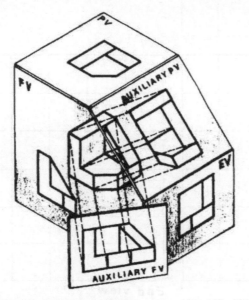

Fig. 5.57 Principal and auxiliary planes

When viewing the object directly, at 90°, through the vertical principal plane (VP), the image of the object will correspond to the front view (FV). When this vertical plane is rotated through 90° about a vertical axis, the corresponding view is the end view (EV). Any other position of the vertical plane between the front view and the end view will correspond to an auxiliary view.

Similarly, when the vertical plane is rotated through 90° about a horizontal axis, it becomes the horizontal principal plane (HP) and its corresponding view is the plan view (PV). Again, any intermediate position of the plane is an auxiliary plane with its corresponding auxiliary view.

In general, referring to Fig. 5.57, any view other than the front, plan, or end view is an auxiliary view.

Figure 5.58 shows how the auxiliary view is obtained between the front and end views in third- and first-angle projection.

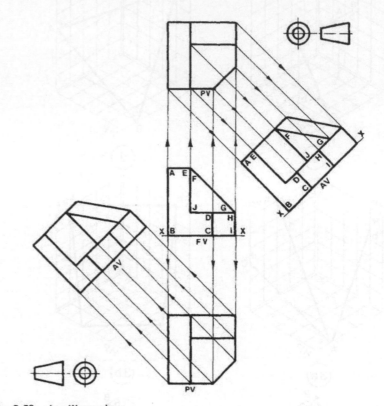

Fig. 5.58 Auxiliary view

To draw the auxiliary view ((Fig. 5.58)
1 Draw the front view (FV) and plan view (PV).
2 Draw the chosen reference line XX in the front view and similarly in the auxiliary view at right angles to the direction of viewing.
3 Transfer all required vertical distances measured from XX in the front view along projectors to the relevant points in the plan view and reflect them to the auxiliary view. Measure the same distances from the reference line XX in the auxiliary view.
4 Join the points so obtained to complete the required auxiliary view (AV).

52

Figure 5.59 shows how the auxiliary view is obtained between the end and plan views in third- and first-angle projection.

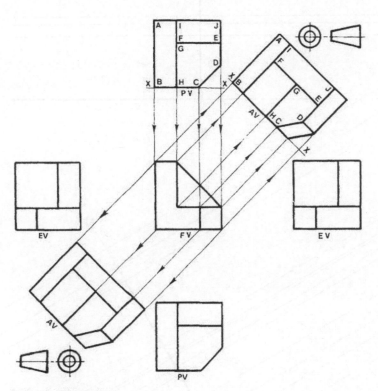

Fig. 5.59 Auxiliary view

To draw the auxiliary view (Fig 5.59)
1 Draw the front and plan views.
2 Draw the chosen reference line XX in the plan view and similarly in the auxiliary view at right angles to the direction of viewing.
3 Transfer all required vertical distances measured from XX in the plan view along projectors to the relevant points in the front view and reflect them to the auxiliary view. Measure the same distances from the reference line XX in the auxiliary view.
4 Join the points so obtained to complete the required auxiliary view (AV).

5.12 Test questions
1 Draw an auxiliary view of each object shown in Fig. 5.60, looking in the direction of arrow Y. To clarify the drawing of auxiliary views, some of the points are indicated with capital letters. The letters with dashes refer only to points on the base of objects.

Note that, in order to draw a surface or edge in the auxiliary view, all projectors must be reflected from that particular surface edge in the front or plan view before the measurements can be taken, as shown in Figure 4 below.

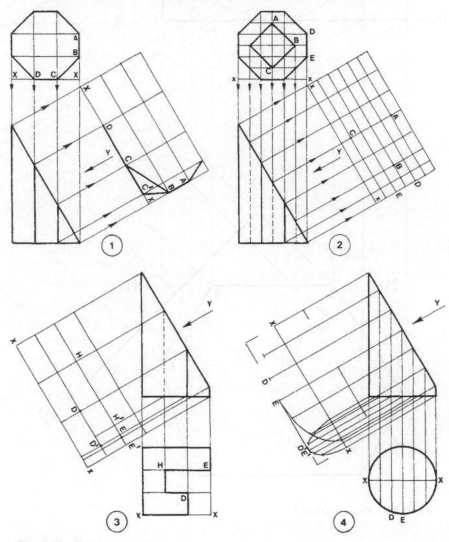

Fig. 5.60 Test question 1

53

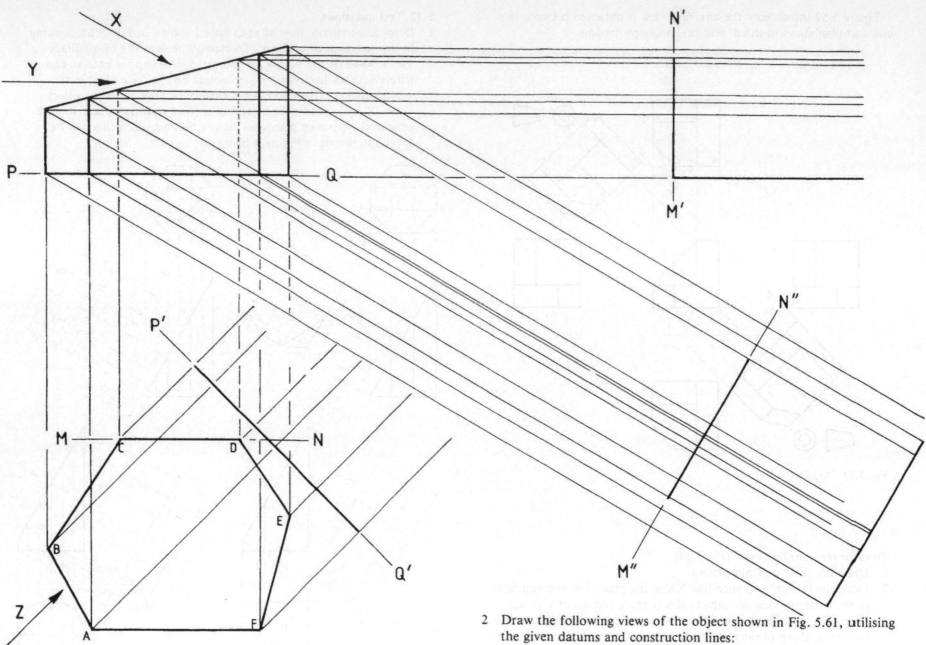

Fig. 5.61 Test question 2

2 Draw the following views of the object shown in Fig. 5.61, utilising the given datums and construction lines:
(a) an end view from N′M′ in the direction Y,
(b) an auxiliary front view from P′Q′ in the direction Z,
(c) an auxiliary plan view from N″M″ in the direction X. Tracing paper may be used.

3 Draw an auxiliary view of each component shown in Fig. 5.62 looking in the direction of the arrow normal to the inclined surface. Components 1 to 4 are in first-angle projection and 4 to 6 in third-angle. Tracing paper may be used.

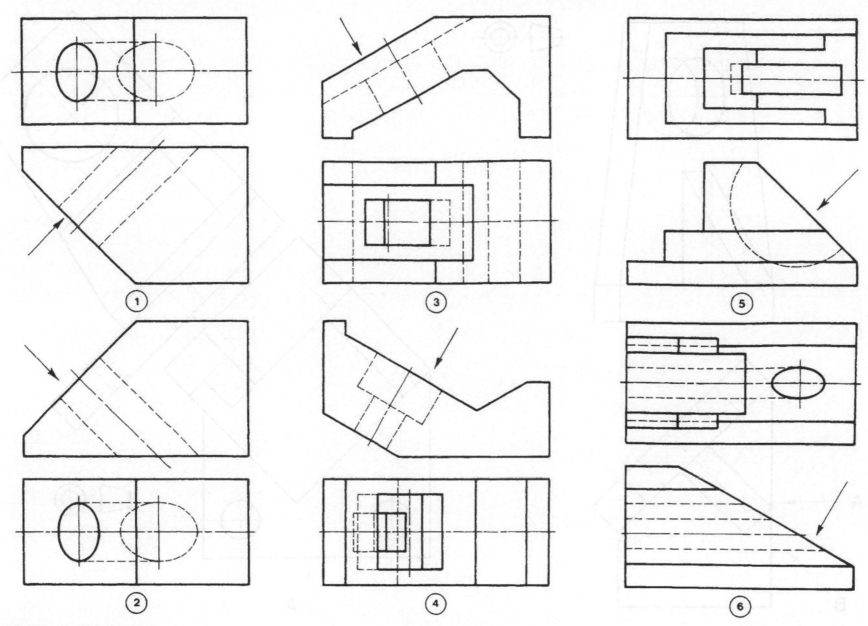

Fig. 5.62   Test question 3

4   For the component shown in Fig. 5.63, draw an auxiliary view looking in the direction of the arrow B and an end view looking in the direction of the arrow A. Tracing paper may be used.

5   Draw an auxiliary view of the component shown in Fig. 5.64 looking in the direction of arrow A. Tracing paper may be used.

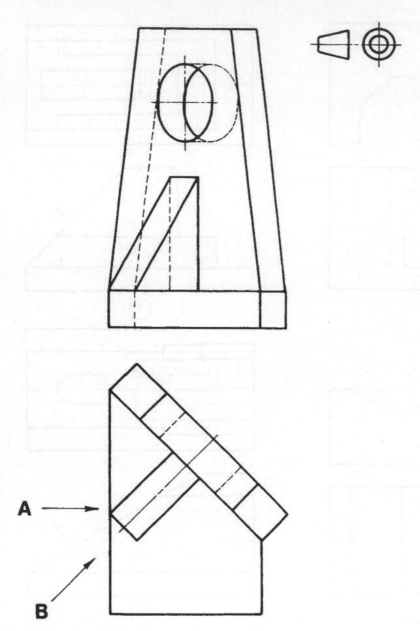

Fig. 5.63   Test question 4

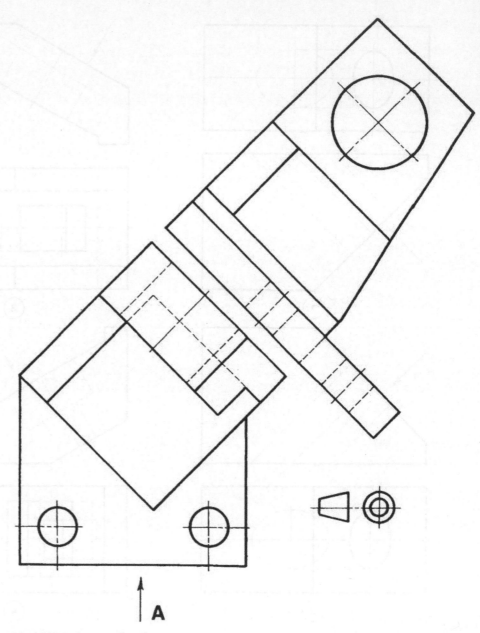

Fig. 5.64   Test question 5

56

6 Dimensioning

An engineering drawing conveys information in two ways:

(a) by a pictorial or orthographic view of the object,
(b) by *instructions* in the form of given sizes or dimensions and notes specifying the manufacturing processes and materials.

Dimensions can be considered to be of two types:

(i) those which define the size and shape of an object or feature – called *size dimensions*;
(ii) those which specify the relative positions of various features – called *location dimensions*.

In addition, dimensions can be put into three groups relative to the function of a product: functional dimensions, non-functional dimensions, and auxiliary dimensions.

6.1 Functional dimensions

These dimensions directly affect the function or working of a product and may be of the size or location types.

Functional dimensions should be based on the function of the component and they can also show the method of locating the component in its appropriate assembly, thus ensuring its correct working within the whole product.

A *datum* is a reference line on the drawing from which a component is dimensioned. In practice, a datum is any functional surface or axis used for manufacture, inspection, location, or assembly purposes. To ensure the required accuracy of measurement, the datum surfaces are machined to the required degree of finish.

The main function of the components in the assembly shown in Fig. 6.1 is to support a shaft. Hence the functional location dimension, F_1, is between the hole centre and the datum mating face.

The fixing holes must be positioned in relation to the shaft and the shoulder on the mating face of the table. The functional dimension F_2 is used for this purpose.

The same dimension F_2 will apply to the bracket and the table. Also, the centre lines of the holes must be positioned relative to the vertical datum using the functional dimensions F_3 and F_4.

Functional dimensions are used for production and inspection purposes. They should always be toleranced.

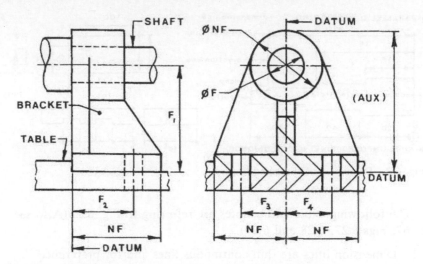

Fig. 6.1 Functional, non-functional, and auxiliary dimensions

6.2 Non-functional dimensions

Non-functional dimensions, NF in Fig. 6.1, are those dimensions which are used for production purposes but which do not directly affect the function or working of a product. Non-functional dimensions usually are not inspected.

6.3 Auxiliary dimensions

Auxiliary dimensions, (AUX) in Fig. 6.1, are given for information only. They are not used for production or inspection purposes, they should not be toleranced, and they should always be inserted in parentheses (brackets).

Auxiliary dimensions are *redundant* dimensions which provide useful information but do not govern acceptance of the product.

6.4 Principles of dimensioning

Dimensions are normally expressed in millimetres. The decimal point should be bold and placed on the base line of the numbers. Dimensions of less than unity should be preceded by zero, e.g. 0.6 mm.

Each dimension should appear only once – it should not be repeated on other views.

Dimensions relative to a particular feature should be placed in one view, which shows the relevant features most clearly, rather than spread over several views.

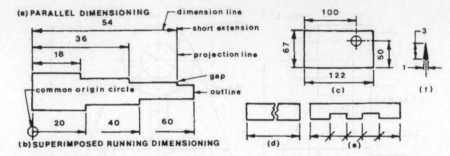

Fig. 6.2 Principles of dimensioning

The following explanatory notes are referring to Fig. 6.2. (Also see p. 67, Figs 6.27, 6.28 and 6.29).

1 Dimension lines are thin continuous lines and for preference should be placed outside the component, Fig. 6.2(a).
2 Projection lines are thin continuous lines projected from outlines. The crossing of projection and dimension lines with other lines should be kept to a minimum.
3 A small gap should be left between the outline and the start of a projection line. The projection lines should continue slightly beyond the dimension line forming a short extension.
4 In parallel dimensioning, dimension lines should end in arrowheads and must touch the extension lines. Arrowheads should be about 3 mm long and 1 mm wide approximately, triangular and filled in, Fig. 6.2(f).
5 Dimension lines should be well spaced, equidistant, and placed outside the outlines of the component, Fig. 6.2(a).
6 Smaller dimensions should be placed nearest to the outlines of the component and larger dimensions outside smaller dimensions.
7 Numerals or letters should preferably be placed centrally above and clear of their dimension lines.
8 In superimposed running dimensioning thin line common origin circles of 3 mm minimum diameter and arrowheads minimum 3 mm long should be used as terminals on dimension lines, Fig. 6.2(b), also p. 67, 6.30.
9 A centre line, outline, or projection line should never be used as a dimension line, but a centre line may be used as a projection line, Fig. 6.2(c).
10 Numerals should be placed so that they may be read from the bottom or from the right-hand side of the drawing, Fig. 6.2(c).
11 Dimension lines should be drawn unbroken for interrupted features, Fig. 6.2(d).

12 Where there is severe space limitation oblique strokes of minimum 3 mm length may be drawn at 45° instead of arrowheads, Fig. 6.2(e).

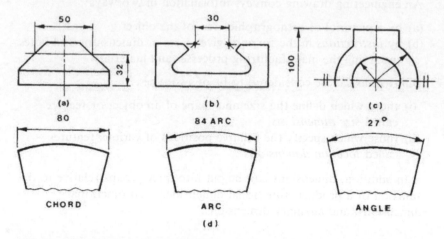

Fig. 6.3 Principles of dimensioning

1 Projection lines referring to points on surfaces, Fig. 6.3(a) should touch the points or referring to imaginary points of intersection, Fig. 6.3(b) should pass through the points. For clarity these points may be indicated by a small dot, as shown in Fig. 6.3(b).
2 Partial views of symmetrical parts should be dimensioned with dimension lines extending a short distance beyond the axis of symmetry and the sector arrowheads omitted, as shown in Fig. 6.3(c).
3 Projection lines and dimension lines for chords, arcs and angles should be drawn as shown in Fig. 6.3(d).

Leaders (Fig. 6.4)
Leaders are thin continuous lines indicating the outlines or surfaces to which relevant dimensions or notes apply.
Leaders end in arrowheads when touching and stopping on a line (Fig. 6.4(a)) but in dots when crossing the line (Fig. 6.4(b)), or may join centrally, a dimension line (Fig. 6.4(c)).
Leaders touching lines should be nearly at right angles to those lines (Fig. 6.4(d)) and when pointing to an arc, should align with its centre (Fig. 6.4(e)).

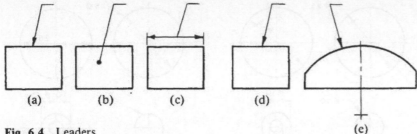

Fig. 6.4 Leaders

Dimensioning common features

When notes or dimensions refer to *repeated features*, a long or intersecting leader should not be used. The dimensions should be repeated as in Fig. 6.5(a) or letter symbols should be used as Fig. 6.6(a).

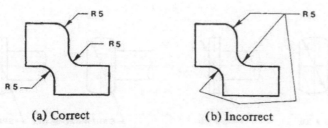

(a) Correct   (b) Incorrect

Fig. 6.5 Dimensioning repeated features

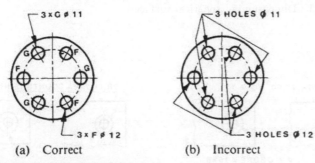

(a) Correct   (b) Incorrect

Fig. 6.6 Dimensioning repeated features

All *radius* dimension lines should pass through or be in line with the centre of the arcs, and they should have only one arrowhead, touching that arc.

Fig. 6.7(a) shows how radii of arcs are dimensioned with centres located and not located.

Dimensioning of *angular positions* of holes on a pitch circle is shown in Fig. 6.8(a).

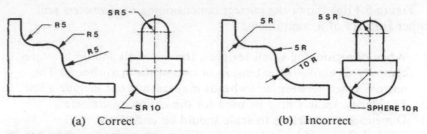

(a) Correct   (b) Incorrect

Fig. 6.7 Dimensioning radii

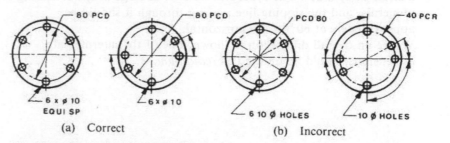

(a) Correct   (b) Incorrect

Fig. 6.8 Dimensioning angular positions

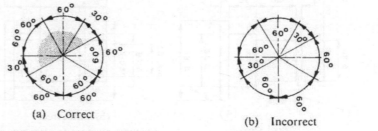

(a) Correct   (b) Incorrect

Fig. 6.9 Placing angular dimensions

Placing of *angular dimensions* is shown in Fig. 6.9(a). The shaded area is used only to highlight the position of the corresponding numerals.

Chamfers of 45° and chamfers at angles other than 45° are dimensioned as shown in Fig. 6.10(a).

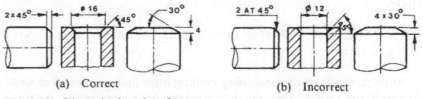

(a) Correct   (b) Incorrect

Fig. 6.10 Dimensioning chamfers

Figure 6.11(a) shows the correct dimensioning of *diameters* and other features of a component:

1 When dimensioning small features, the numerals should be placed centrally or above the extension of one of the arrowheads. The narrow space between arrowheads may or may not include a line.
2 For clarity, leaders may be used for dimensioning diameters.
3 Dimensions not drawn to scale should be underlined.
4 Symbol Ø should be in front of a dimension giving the diameter of a circle or cylinder. The symbol should be as large as the following numerals, and the sloping line passing through it should be approximately at 60° to the horizontal.
5 Where an overall dimension is shown, one of the intermediate dimensions should be omitted as *redundant*.

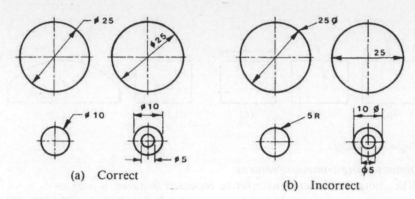

Fig. 6.12   Dimensioning circles

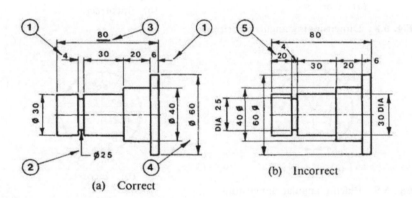

Fig. 6.11   Dimensioning diamaters

Fig. 6.12(a) shows four methods of dimensioning *circles*. When a leader touches a circle, it should be in line with the centre of that circle.

The diameter of a *spherical surface* should be dimensioned as shown in Fig. 6.13(a).

Some methods of dimensioning countersinks and counterbores and of specifying tapered features on a drawing are shown in Fig. 6.14(a).

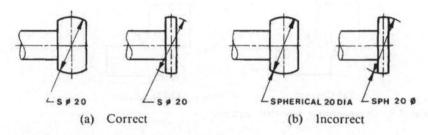

Fig. 6.13   Dimensioning spherical surfaces

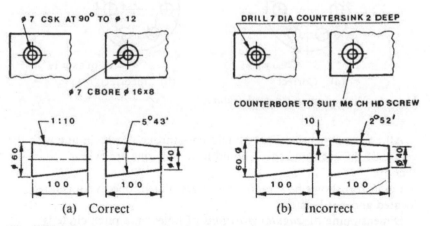

Fig. 6.14   Dimensioning countersinks, counterbores, and tapered feature

6.5 Dimensioning for different purposes

Every drawing must include as many dimensions as are necessary for the complete definition of the component, describing the size, shape, and positioning of all relevant features.

Sometimes a component requires several drawings for different stages of manufacture, with corresponding dimensions.

6.6 Dimensioning for primary production

For primary production – which may include casting, forging, or fabrication – the drawing should have all dimensions necessary for shaping the object at that stage of production.

Sand casting

In sand casting, a cavity having the shape of the required component is formed in a box of sand by a wooden mould called a pattern. Molten metal is poured into the cavity and is allowed to cool. The casting is then removed and is ready to be machined. In sand casting, depending on the brittleness, castability, and strength of the materials, the casting usually has strengthening ribs or webs and blending fillets. When dimensioning, Fig. 6.15, an allowance for shrinkage and warping has usually to be taken into account.

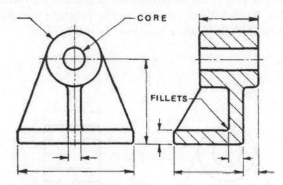

Fig. 6.15   Dimensioning for sand casting

Die casting

In die casting, the molten metal is forced into a mould under pressure, making it possible to get the surfaces more true and smooth and the dimensions more accurate than in sand casting, as there is less shrinkage and warping.

Drop forging

In drop forging, the metal is heated and then hammered into the desired shape, using a die.

In dimensioning, Fig. 6.16, all corners of the component are rounded to allow correct metal formation, and the sides are usually tapered to facilitate the removal of the component from the die after forging.

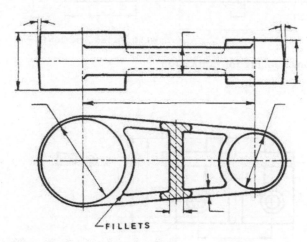

Fig. 6.16   Dimensioning for drop forging

Fabrication

In fabrication, a welded construction is used for small quantities of components which usually cannot be cast because of strength requirements and cannot readily be forged because of their shape or size.

Fabricated components are built of several different pieces of steel cut to shape and then welded together. For dimensioning, Fig. 6.17, the shapes and relative positions of all separate pieces have to be taken into consideration, and all welding instructions must be clearly stated (see pages 109–110).

6.7 Dimensioning for secondary production

For the purposes of secondary production, which may include any machining processes, the drawing should have only those dimensions necessary for machining the required surfaces, as in Fig. 6.18.

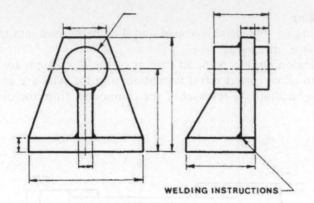

Fig. 6.17 Dimensioning for fabrication

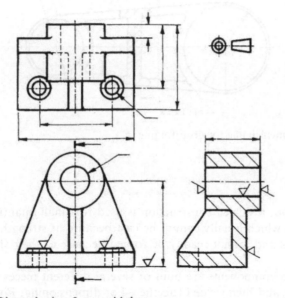

Fig. 6.18 Dimensioning for machining

Inspection

For the purposes of inspection, which is to ensure that components are within the dimensional limits laid down, the drawing will include all dimensions necessary to indicate the maximum and minimum limits of the machined sizes and the roughness of surfaces if required.

Rapid and economic inspection of limits is carried out by means of different types of gauge. The principle of limit gauging is that work is acceptable if the 'GO' section of a limit gap gauge passes over the work and the 'NO GO' section does not pass, Fig. 6.19.

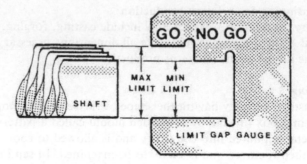

Fig. 6.19 Principle of limit gauging

Functional dimensioning

For functional purposes, the drawing should include all functional dimensions that have a direct bearing on the function and working of the components or their relative location, as in Fig. 6.20.

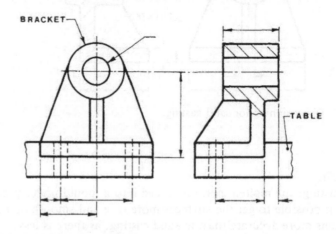

Fig. 6.20 Functional dimensions

Detail drawings should contain sufficient information for the production department to make the components. This should include a complete set of dimensions and tolerances, see Fig. 6.21(a).

For *assembly* purposes the drawing may include overall and fitting dimensions. Sometimes the functional dimensions are included in this type of drawing, as shown in Fig. 6.21(b).

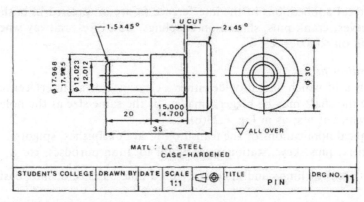

Fig. 6.21(a) Complete dimensioning of a single part

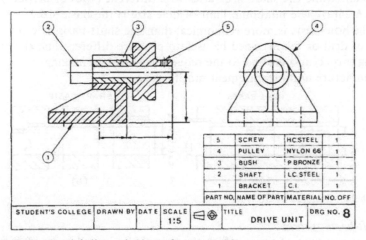

Fig. 6.21(b) Partial dimensioning of an assembly

6.8 Tolerances

It is impossible to manufacture a component to the exact design size, this size being a numerical value of length. To overcome this difficulty,

a tolerance is permitted or 'tolerated' which is the amount of deviation from the given basic design size, or the margin of error, allowable to accommodate reasonable inaccuracy in manufacturing.

On a drawing, this tolerance is indicated by the maximum and minimum permitted sizes, which are called *limits of size*, as shown in Fig. 6.22. Tolerances should be stated only where the required accuracy is essential.

There are three principle considerations in establishing reasonable tolerances: function, interchangeability, and cost.

Function

The tolerances must be consistent with the design and function or working of the component. If a shaft is designed to rotate in a bearing, the tolerances must ensure that the shaft functions properly and is never larger than the hole of the bearing.

Cost

The expense of manufacture varies directly with the accuracy required; hence tolerances should be as large as the design will permit without affecting the function of the component. Large tolerances introduce savings in labour, use of tools, and running the machines.

Interchangeability

Mass production often depends on producing certain parts on different machines to predetermined tolerances to ensure that all assembled parts will always fit together to form properly operating units. This can be controlled and communicated to the workshops only through the detail working drawings, and interchangeability can be ensured through inspection utilising an efficient system of gauging.

In general, limits of size should be applied only if the specified accuracy is essential for the efficient functioning of the components or to facilitate their interchangeability.

Figure 6.22 shows the elements of an interchangeable system where two parts fit together – an external part (a shaft) and an internal feature of a part (a hole).

Basic size is usually a theoretical design size to which all limits of size are referred.

It is the same for both members of the fit.

Deviation is the algebraic difference between an actual size, obtainable by measurement, and the corresponding basic size.

Upper deviation is the algebraic difference between the maximum limit of size and the corresponding basic size.

Lower deviation is the algebraic difference between the minimum limit of size and the corresponding basic size.

Zero line is the line of zero deviation and represents the basic size.

Maximum limit of size is the maximum size permitted for a feature.

Minimum limit of size is the minimum size permitted for a feature.

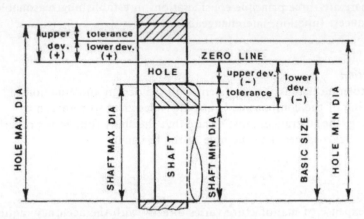

Fig. 6.22 Elements of an interchangeable system

A designer must ensure that an assembly of mating components will function correctly, thus all parts must fit together in the required manner.

A particular fit will depend solely on the prescribed maximum and minimum limits of size of the two separate components which are to be assembled. Engineering fits can be divided into three main types: clearance fits, interference fits, and transition fits.

Clearance fit

This is a fit which provides a clearance; hence the shaft is always smaller than the hole into which it fits, as in Fig. 6.23(a). *Clearance* is the positive difference between the sizes of the hole and the shaft.

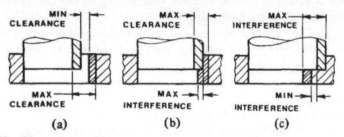

Fig. 6.23 Three types of fit

Typical applications of the clearance fit are on rotating shafts, loose pulleys, fast pulleys, bearings, cross-head slides, etc.

Interference fit

This is a fit which always provides an interference; hence the shaft is always bigger than the hole into which it fits, as in Fig. 6.23(c). *Interference* is the negative difference between the sizes of the hole and the shaft.

Typical applications of the interference fit are on pressed-in bushes or sleeves, crank pins, shrunk-on couplings, iron tyres, railway wheels shrunk on to axles, etc.

Transition fit

This is a fit which may provide either a clearance or an interference; hence the shaft may be bigger, smaller, or the same size as the hole into which it fits, as in Fig. 6.23(b).

Typical applications of the transition fit are on bushes, spigots, fasteners, pins, keys, stationary parts for location purposes, etc.

A system of limits and fits may be on a hole basis or a shaft basis.

Hole-basis system

This is a system of fits in which the basic diameter of the hole is constant while the shaft size varies with different types of fit, see Fig. 6.24(a). The minimum limit of hole size is the basic size.

The hole-basis is more economical than the shaft-basis as only one size of drill or reamer need be used to produce different fits, the shafts being turned and ground to the required sizes, thus making manufacture and measurement much easier.

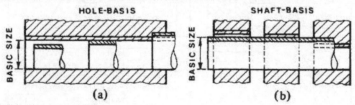

Fig. 6.24 Systems of fit

Shaft-basis system

This is a system of fits in which the hole size is varied to produce the required type of fit, with the basic diameter of the shaft being constant. The maximum limit of the shaft is the basic size, see Fig. 6.24(b).

This system tends to be less economical, as a series of drills is required. It is usually adopted where a single driving shaft accommodates a number of pulleys, bearings, collars, couplings, etc.

SELECTED ISO FITS—HOLE BASIS

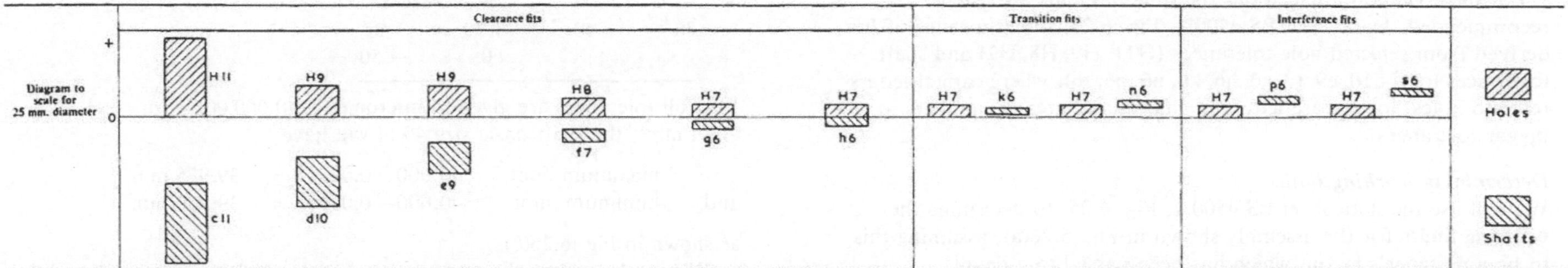

Diagram to scale for 25 mm. diameter — Holes / Shafts

Clearance fits: H11/c11, H9/d10, H9/e9, H8/f7, H7/g6, H7/h6
Transition fits: H7/k6, H7/n6
Interference fits: H7/p6, H7/s6

All tolerance values in 0·001 mm.

Over (mm)	To (mm)	H11	c11	H9	d10	H9	e9	H8	f7	H7	g6	H7	h6	H7	k6	H7	n6	H7	p6	H7	s6	Over (mm)	To (mm)
—	3	+60/0	-60/-120	+25/0	-20/-60	+25/0	-14/-39	+14/0	-6/-16	+10/0	-2/-8	+10/0	-6/0	+10/0	+6/0	+10/0	+10/+4	+10/0	+12/+6	+10/0	+20/+14	—	3
3	6	+75/0	-70/-145	+30/0	-30/-78	+30/0	-20/-50	+18/0	-10/-22	+12/0	-4/-12	+12/0	-8/0	+12/0	+9/+1	+12/0	+16/+8	+12/0	+20/+12	+12/0	+27/+19	3	6
6	10	+90/0	-80/-170	+36/0	-40/-98	+36/0	-25/-61	+22/0	-13/-28	+15/0	-5/-14	+15/0	-9/0	+15/0	+10/+1	+15/0	+19/+10	+15/0	+24/+15	+15/0	+32/+23	6	10
10	18	+110/0	-95/-205	+43/0	-50/-120	+43/0	-32/-75	+27/0	-16/-34	+18/0	-6/-17	+18/0	-11/0	+18/0	+12/+1	+18/0	+23/+12	+18/0	+29/+18	+18/0	+39/+28	10	18
18	30	+130/0	-110/-240	+52/0	-65/-149	+52/0	-40/-92	+33/0	-20/-41	+21/0	-7/-20	+21/0	-13/0	+21/0	+15/+2	+21/0	+28/+15	+21/0	+35/+22	+21/0	+48/+35	18	30
30	40	+160/0	-120/-280	+62/0	-80/-180	+62/0	-50/-112	+39/0	-25/-50	+25/0	-9/-25	+25/0	-16/0	+25/0	+18/+2	+25/0	+33/+17	+25/0	+42/+26	+25/0	+59/+43	30	40
40	50	+160/0	-130/-290	+62/0	-80/-180	+62/0	-50/-112	+39/0	-25/-50	+25/0	-9/-25	+25/0	-16/0	+25/0	+18/+2	+25/0	+33/+17	+25/0	+42/+26	+25/0	+59/+43	40	50
50	65	+190/0	-140/-330	+74/0	-100/-220	+74/0	-60/-134	+46/0	-30/-60	+30/0	-10/-29	+30/0	-19/0	+30/0	+21/+2	+30/0	+39/+20	+30/0	+51/+32	+30/0	+72/+53	50	65
65	80	+190/0	-150/-340	+74/0	-100/-220	+74/0	-60/-134	+46/0	-30/-60	+30/0	-10/-29	+30/0	-19/0	+30/0	+21/+2	+30/0	+39/+20	+30/0	+51/+32	+30/0	+78/+59	65	80
80	100	+220/0	-170/-390	+87/0	-120/-260	+87/0	-72/-159	+54/0	-36/-71	+35/0	-12/-34	+35/0	-22/0	+35/0	+25/+3	+35/0	+45/+23	+35/0	+59/+37	+35/0	+93/+71	80	100
100	120	+220/0	-180/-400	+87/0	-120/-260	+87/0	-72/-159	+54/0	-36/-71	+35/0	-12/-34	+35/0	-22/0	+35/0	+25/+3	+35/0	+45/+23	+35/0	+59/+37	+35/0	+101/+79	100	120
120	140	+250/0	-200/-450	+100/0	-145/-305	+100/0	-85/-185	+63/0	-43/-83	+40/0	-14/-39	+40/0	-25/0	+40/0	+28/+3	+40/0	+52/+27	+40/0	+68/+43	+40/0	+117/+92	120	140
140	160	+250/0	-210/-460	+100/0	-145/-305	+100/0	-85/-185	+63/0	-43/-83	+40/0	-14/-39	+40/0	-25/0	+40/0	+28/+3	+40/0	+52/+27	+40/0	+68/+43	+40/0	+125/+100	140	160
160	180	+250/0	-230/-480	+100/0	-145/-305	+100/0	-85/-185	+63/0	-43/-83	+40/0	-14/-39	+40/0	-25/0	+40/0	+28/+3	+40/0	+52/+27	+40/0	+68/+43	+40/0	+133/+108	160	180
180	200	+290/0	-240/-530	+115/0	-170/-355	+115/0	-100/-215	+72/0	-50/-96	+46/0	-15/-44	+46/0	-29/0	+46/0	+33/+4	+46/0	+60/+31	+46/0	+79/+50	+46/0	+151/+122	180	200
200	225	+290/0	-260/-550	+115/0	-170/-355	+115/0	-100/-215	+72/0	-50/-96	+46/0	-15/-44	+46/0	-29/0	+46/0	+33/+4	+46/0	+60/+31	+46/0	+79/+50	+46/0	+159/+130	200	225
225	250	+290/0	-280/-570	+115/0	-170/-355	+115/0	-100/-215	+72/0	-50/-96	+46/0	-15/-44	+46/0	-29/0	+46/0	+33/+4	+46/0	+60/+31	+46/0	+79/+50	+46/0	+169/+140	225	250
250	280	+320/0	-300/-620	+130/0	-190/-400	+130/0	-110/-240	+81/0	-56/-108	+52/0	-17/-49	+52/0	-32/0	+52/0	+36/+4	+52/0	+66/+34	+52/0	+88/+56	+52/0	+190/+158	250	280
280	315	+320/0	-330/-650	+130/0	-190/-400	+130/0	-110/-240	+81/0	-56/-108	+52/0	-17/-49	+52/0	-32/0	+52/0	+36/+4	+52/0	+66/+34	+52/0	+88/+56	+52/0	+202/+170	280	315
315	355	+360/0	-360/-720	+140/0	-210/-440	+140/0	-125/-265	+89/0	-62/-119	+57/0	-18/-54	+57/0	-36/0	+57/0	+40/+4	+57/0	+73/+37	+57/0	+98/+62	+57/0	+226/+190	315	355
355	400	+360/0	-400/-760	+140/0	-210/-440	+140/0	-125/-265	+89/0	-62/-119	+57/0	-18/-54	+57/0	-36/0	+57/0	+40/+4	+57/0	+73/+37	+57/0	+98/+62	+57/0	+244/+208	355	400
400	450	+400/0	-440/-840	+155/0	-230/-480	+155/0	-135/-290	+97/0	-68/-131	+63/0	-20/-60	+63/0	-40/0	+63/0	+45/+5	+63/0	+80/+40	+63/0	+108/+68	+63/0	+272/+232	400	450
450	500	+400/0	-480/-880	+155/0	-230/-480	+155/0	-135/-290	+97/0	-68/-131	+63/0	-20/-60	+63/0	-40/0	+63/0	+45/+5	+63/0	+80/+40	+63/0	+108/+68	+63/0	+292/+252	450	500

Fit names: Slack fit | Loose fit | Easy fit | Normal fit | Close fit | Slide fit | Push fit | Drive fit | Press fit | Force fit

Fig. 6.25 British Standard data sheet BS 4500A; selected ISO fits – hole basis

British Standard BS 4500, *ISO limits and fits*, gives a selection of hole and shaft tolerances to cover a wide range of engineering applications.

For a selected range of fits which is adequate for most practical requirements, the BS 4500A and BS 4500B data sheets give the fits on a hole and shaft basis respectively.

For most general applications the hole-basis fits are usually recommended. Data sheet BS 4500A, Fig. 6.25, shows a range of fits derived from selected hole tolerances (H11, H9, H8, H7) and shaft tolerances (c11, d10, e9, f7, g6, h6, k6, n6, p6, s6), where capital letters refer to holes, lower-case letters to shafts, and greater numbers to bigger tolerances.

Determining working limits

We will use the data sheet BS 4500A, Fig. 6.25, to determine the working limits for the assembly shown in Fig. 6.26(a), assuming this to be a designer's layout which has been passed to a detail draughtsman for preparation of the working drawings of the components.

In order to decide on a desirable fit, we must consider the function of the assembly. The shaft is going to rotate in the bush; hence a clearance fit is required, allowing sufficient space for lubricant, but not so much as to cause wobbling of the shaft. The most suitable fit will be H8/f7, as shown in Fig. 6.26(b).

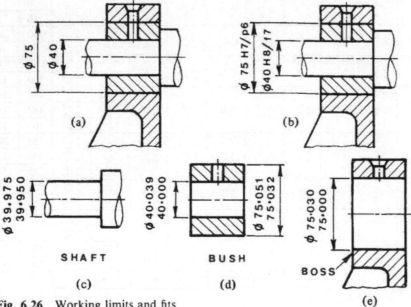

Fig. 6.26 Working limits and fits

We now locate 40 mm in the column of nominal sizes in the data sheet BS 4500A in Fig. 6.25, remembering that column 'Over' means 'over but excluding' and 'To' means 'to and including'.

The required tolerances are shown below:

Over	To	H8	f7
30	40	+ 39	− 25
		0	− 50

As all tolerances are given in micrometres (0.000 001 m or 0.001 mm), the shaft basic size 40 f7 will have

	maximum limit	40.000 − 0.025	=	39.975 mm
and	minimum limit	40.000 − 0.050	=	39.950 mm

as shown in Fig. 6.26(c).

(When tolerancing, the same number of decimal places must be used for both limits.)

The bush hole basic size 40 H8 will have

	maximum limit	40.000 + 0.039	=	40.039 mm
and	minimum limit	40.000 + 0	=	40.000 mm

as shown in Fig. 6.26(d).

Assuming that the bush is going to be pressed into the bracket boss, then an interference fit will be appropriate. The h7/p6 fit seems to be most suitable, as shown in Fig. 6.26(b).

Now locate 75 mm in the column of nominal sizes in the data sheet BS 4500A, Fig. 6.25:

Over	To	H7	p6
65	80	+ 30	+ 51
		0	+ 32

The bush outside-diameter basic size 75 p6 will have

	maximum limit	75.000 + 0.051	=	75.051 mm
and	minimum limit	75.000 + 0.032	=	75.032 mm

as shown in Fig. 6.26(d).

The bracket boss basic size 75 H7 will have

	maximum limit	75.000 + 0.030	=	75.030 mm
and	minimum limit	75.000 + 0	=	75.000 mm

as shown in Fig. 6.26(e).

Dimensioning tolerances

A toleranced drawing of a rectangular component is shown in Fig. 6.27(a) and how it is interpreted is shown in Fig. 6.27(b). The tolerance zones of 0.6 mm shown are the differences between the maximum and minimum limits.

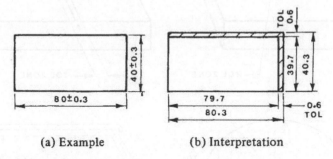

(a) Example   (b) Interpretation

Fig. 6.27   Tolerancing a component

Tolerance limits between centres of holes can be indicated either by *chain dimensioning*, as in Fig. 6.28(a) or by *parallel dimensioning* from a common datum as shown in Fig. 6.29(a).

The use of chain dimensions results in an accumulation of tolerances between the holes and the edge of the plate, and this may endanger the functional requirements, as shown in Fig. 6.28(b).

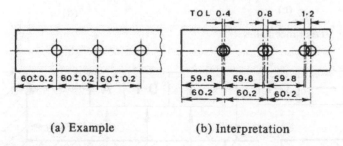

(a) Example   (b) Interpretation

Fig. 6.28   Chain dimensioning

Parallel dimensioning from a common datum on the component prevents this accumulation of tolerances, as each hole is toleranced directly from the datum, as shown in Fig. 6.29(b).

Chain and parallel dimensioning may be combined on the same drawing.

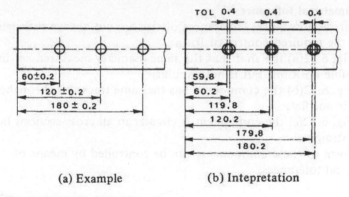

(a) Example   (b) Intepretation

Fig. 6.29   Parallel dimensioning

Superimposed running method of dimensioning (Fig. 6.30) is simplified parallel dimensioning and may be used where space is limited.

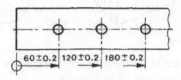

Fig. 6.30   Superimposed running dimensioning

When tolerancing an individual linear dimension, the method of specifying directly maximum and minimum limits of size is preferable. The larger limit should be given first, and the same number of decimal places should be indicated for both limits.

Four correct methods of tolerancing are shown in Fig. 6.31.

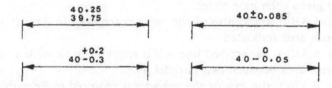

Fig. 6.31   Methods of tolerancing linear dimensions

6.9 Geometrical tolerances

In certain circumstances, tolerances of size are not always sufficient to provide the required control of form:

(a) in Fig. 6.32(a) the shaft has the same diameter measurement in all possible positions but is not circular;

(b) in Fig. 6.32(b) the component has the same thickness throughout but is not flat;

(c) in Fig. 6.32(c) the component is circular in all cross-sections but is not straight.

The form of these components can be controlled by means of geometrical tolerances.

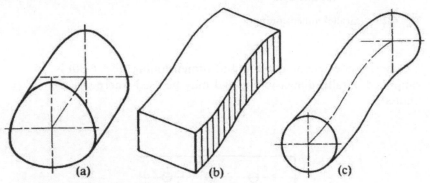

Fig. 6.32 Errors of form

Geometrical tolerance is defined as the maximum permissible overall variation of *form* or *position* of a feature.

Geometrical tolerances are used

(i) to specify the required accuracy in controlling the form of a feature,

(ii) to ensure correct functional positioning of a feature,

(iii) to ensure the interchangeability of components, and

(iv) to facilitate the assembly of mating components.

The *tolerance zone* is an imaginary area or volume within which the controlled feature of the manufactured component must be completely contained.

Figure 6.33 shows four components which, after being inspected, will be passed as correctly manufactured if the features controlled lie within the given tolerance zones.

(a) In Fig. 6.33(a), the centre of the circle is required to lie within the tolerance area indicated.

(b) In Fig. 6.33(b), the scribed line AB is required to lie within the tolerance area between two parallel lines.

(c) In Fig. 6.33(c), the axis of the cylinder is required to lie within the tolerance indicated.

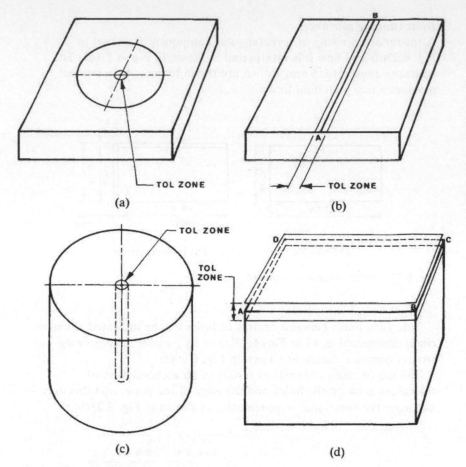

Fig. 6.33 Tolerance zones

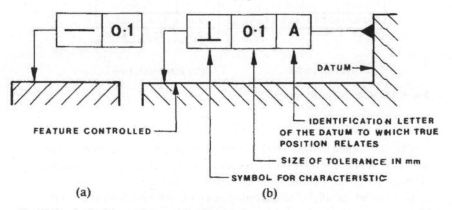

Fig. 6.34 Indication of geometrical tolerances

(d) In Fig. 6.33(d), the surface ABCD is required to lie within the tolerance volume.

6.10 Indicating geometrical tolerances on drawings

To eliminate the need for descriptive notes, geometrical tolerances are indicated on drawings by symbols, tolerances, and datums, all contained in compartments of a rectangular frame, as shown in Fig. 6.34.

Type of tolerance and tolerance symbols

Table 6.1 Symbols for toleranced characteristics

	Type of tolerance	Characteristic to be toleranced	Symbol
For single features	Form	Straightness	⎯
		Flatness	▱
		Roundness	○
		Cylindricity	⌭
		Profile of a line	⌒
		Profile of a surface	⌓
For related features	Attitude	Parallelism	∥
		Squareness	⊥
		Angularity	∠
	Location	Position	⌖
		Concentricity	◎
		Symmetry	≡
	Composite	run-out	⤢

Form tolerance specifies the required geometric shape of a single feature.

Attitude tolerance specifies the required orientation of a feature relative to a datum.

Location tolerance specifies the required position of a feature relative to a datum.

Run-out tolerance is defined in terms of measurement of the maximum overall variation of the surface or the end face of a component during rotation of the component about a specified datum axis, or by rotating it when supported by two specified datum points.

Maximum material condition, MMC

Engineering drawings are usually dimensioned and toleranced to control mating parts at their worst conditions of fit for assembly, i.e. the largest shaft and the smallest hole condition. This condition is called the maximum material condition, MCC, and is denoted by the symbol on drawings, as shown below.

6.11 Advantages of using geometrical tolerances

1 Geometrical tolerances convey very briefly and precisely the complete geometrical requirements on engineering drawings.
2 The use of symbols and boxes eliminates the need for lengthy descriptive notes and corresponding dimensions; therefore the drawings are much clearer to read.
3 The symbols used are internationally recommended; hence the language barrier is minimised and misunderstanding is eliminated.
4 One type of geometrical tolerance can control another form. For instance, squareness can control flatness and straightness.

6.12 Feature controlled

The feature controlled by a geometrical tolerance is indicated by an arrowhead at the end of a leader line from the tolerance frame. When the tolerance refers to an outline of a feature or to a surface

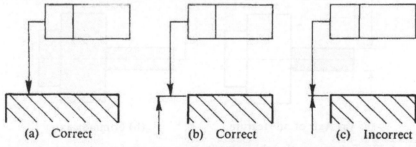

Fig. 6.35 Indication of feature controlled (outline or surface only)

69

represented by an outline, the arrowhead may touch either the outline, as shown in Fig. 6.35(a), or an extension line from the outline, as shown in Fig. 6.35(b), but *not* at a dimension line as shown in Fig. 6.35(c).

If the tolerance refers to the axis or median (central) plane of a single-feature part, the arrowhead may touch either the axis or median plane itself, Fig. 6.36(a), or the dimension line relevant to the feature whose axis is concerned, Fig. 6.36(b).

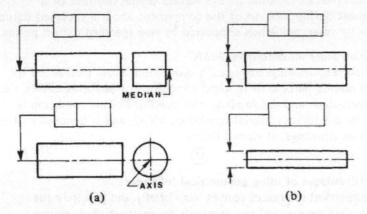

Fig. 6.36 Tolerances of a single-feature part

If the tolerance refers to the axis or median plane of only one feature of a multi-feature part, the arrowhead touches the dimension line relevant to the feature whose axis is concerned, as shown in Fig. 6.37(a).

If the tolerance refers to the common axis or median plane of a number of features, the arrowhead touches the axis or median plane, as shown in Fig. 6.37(b).

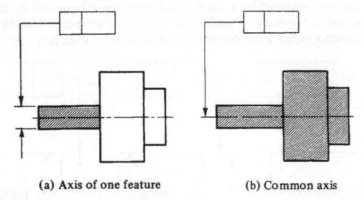

(a) Axis of one feature   (b) Common axis

Fig. 6.37 Tolerances of multi-feature part

6.13 Datum features

A datum may be real or imaginary. It can be a reference plane, surface, or axis used for measuring, location, or inspection purposes.

The datum feature is indicated by a solid equilateral triangle at the end of a leader line from the tolerance frame.

When the datum feature is an outline or a surface represented by an outline, the triangle may be positioned either on the outline or on an extension line from the outline (but not at a dimension line), as shown in Fig. 6.38.

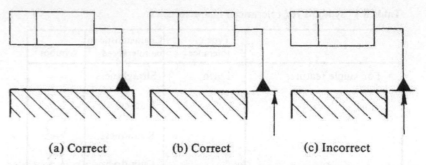

(a) Correct   (b) Correct   (c) Incorrect

Fig. 6.38 Indication of datum feature (outline or surface only)

If the datum feature is the axis or median plane of a single-feature part, the triangle may be positioned either on the axis or median plane itself or on a projection line at the dimension line relevant to the feature whose axis is concerned, as shown in Fig. 6.39(a).

If the datum feature is the axis or median plane of a particular feature of a multi-feature part, the triangle is positioned on a projection line at the dimension line relevant to the feature whose axis is concerned, as shown in Fig. 6.39(b).

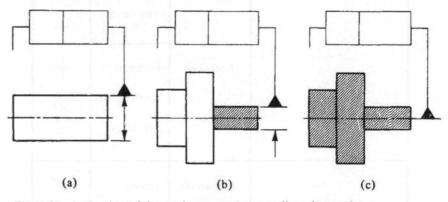

(a)   (b)   (c)

Fig. 6.39 Indication of datum feature (axis or median plane only)

If the datum feature is the common axis or median plane of a number of features, the triangle is positioned on the axis or median plane as shown in Fig. 6.39(c).

If the datum feature cannot be clearly and simply connected to the tolerance frame, then a capital letter is connected to the datum feature and is referred to in a separate part of the tolerance frame as shown in Fig. 6.40.

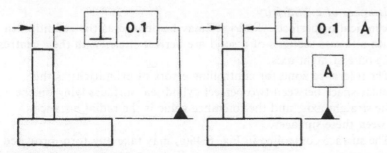

Fig. 6.40 Datum reference

6.14 Boxed dimensions
Dimensions enclosed in a box, e.g. 30 ⌀75 45° EQUI SP define the true position or exact location of a feature on a component.

6.15 General principles of geometrical tolerancing
1 Geometrical tolerances apply to the whole length or surface of the feature unless stated or indicated otherwise. Figure 6.41 shows how a geometrical tolerance is limited to a particular part of the feature.
2 The use of geometrical tolerances does not imply the use of any particular method of production or inspection.
3 A line or surface of a feature controlled by geometrical tolerances may be of any form and may take any position provided it remains within its tolerance zone.

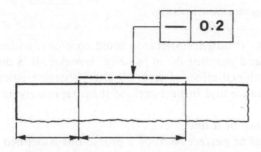

Fig. 6.41 Tolerancing of a particular part of a feature

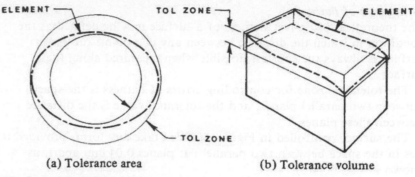

(a) Tolerance area   (b) Tolerance volume

Fig. 6.42 Tolerance zones

Figure 6.42(a) shows a line within its tolerance area and Fig. 6.42(b) shows a surface within its tolerance volume.

6.16 Tolerances of form for single features

Tolerances of straightness
The theoretical or perfect straightness of a line on a surface may be defined as the condition in which the distance between any two points on that line is always the shortest possible when measured along the line.

The tolerance zone for controlling errors of straightness is the area between two parallel lines, and the tolerance value is the distance between these lines.

The line on the surface of the feature in Fig. 6.43(a) can take any form, provided it lies in an axial plane between two parallel straight lines 0.02 mm apart, as shown in Fig. 6.43(b).

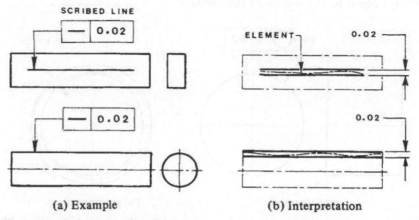

(a) Example   (b) Interpretation

Fig. 6.43 Tolerances of straightness

Tolerances of flatness

The theoretical or perfect flatness of a surface may be defined as the condition in which the distance between any two points on that surface is always the shortest possible when measured along that surface.

The tolerance zone for controlling errors of flatness is the space between two parallel planes, and the tolerance value is the distance between these planes.

The surface controlled in Fig. 6.44(a) can take any form, provided it lies in the space between two parallel flat planes 0.04 mm apart, as shown in Fig. 6.44(b).

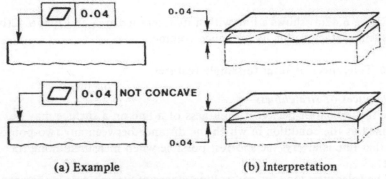

(a) Example   (b) Interpretation

Fig. 6.44 Tolerances of flatness

Tolerances of roundness

The theoretical or perfect roundness of a surface may be defined as the condition in which the surface has the form of a perfect circle, i.e. the distance between any point on the circumference and the centre is always equal to the radius of the circle.

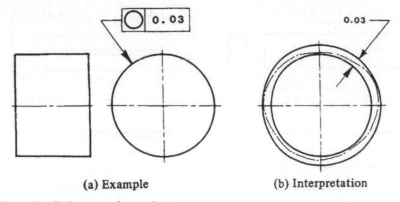

(a) Example   (b) Interpretation

Fig. 6.45 Tolerance of roundness

The tolerance zone for controlling errors of roundness is the annular area between two concentric coplanar circles, and the tolerance value is the radial distance between these circles.

In Fig. 6.45(a), the circle controlled, which may represent the periphery at any cross-section perpendicular to the axis, can take any form provided it lies in the space between two concentric circles 0.03 mm radially apart, as shown in Fig. 6.45(b).

Tolerances of cylindricity

Theoretical or perfect cylindricity may be defined as the condition in which all cross sections of a solid are perfect circles with their centres lying on a straight axis.

The tolerance zone for controlling errors of cylindricity is the annular space between two perfect cylindrical surfaces lying on the same straight axis, and the tolerance value is the radial distance between these surfaces.

The surface controlled in Fig. 6.46(a) may take any form provided it lies between two perfect concentric cylinders 0.03 mm apart, as shown in Fig. 6.46(b).

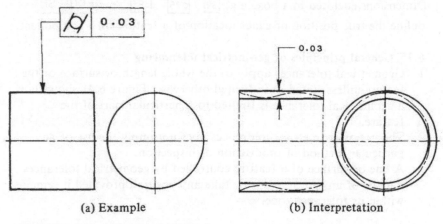

(a) Example   (b) Interpretation

Fig. 6.46 Tolerance of cylindricity

In theory, a cylindricity tolerance could control roundness, straightness, and parallelism; in practice, however, it is difficult to check the combined effect of errors in these characteristics, and it is better to tolerance and inspect each of them separately as required.

Profile tolerance of a line

The theoretical or perfect form of a profile line is defined by boxed dimensions, which locate the true position of any point on that line.

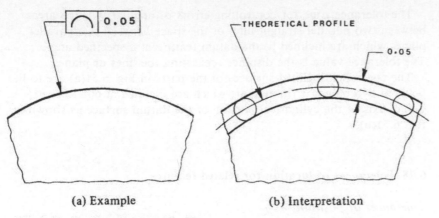

(a) Example (b) Interpretation

Fig. 6.47 Profile tolerance of a line

The tolerance zone has a constant width equal to the tolerance value, normal (at 90°) to the theoretical profile and equally disposed about it.

The tolerance zone is the area between two lines which envelop circles of diameter equal to the tolerance value.

The profile line controlled in Fig. 6.47(a) can take any form provided it lies between two lines 0.05 mm apart, as shown in Fig. 6.47(b).

Profile tolerances of a surface

The theoretical or perfect form of a surface is defined by boxed dimensions which locate the true position of any point on that surface.

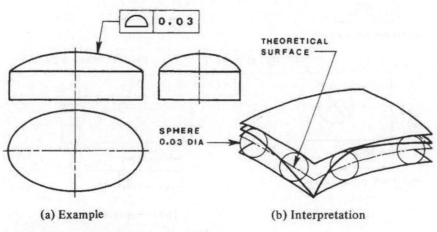

(a) Example (b) Interpretation

Fig. 6.48 Profile tolerance of a surface

The tolerance zone is the space between two surfaces which envelop spheres of diameter equal to the tolerance value with their centres lying on the theoretical surface of the correct geometrical shape.

The curved surface of the part controlled in Fig. 6.48(a) is required to lie between two surfaces as shown in Fig. 6.48(b).

6.17 Tolerances of attitude for related features

Tolerances of parallelism

Theoretical or perfect parallelism may be defined as the condition in which all the perpendicular distances between the line or the surface controlled and the datum feature are always the same.

The tolerance zone for controlling errors of parallelism is the area between two parallel straight lines or the space between two parallel planes which are parallel to the datum feature. The tolerance value is the distance between the lines or planes.

The controlled top surface of the part shown in Fig. 6.49(a) is required to lie between two planes 0.06 mm apart and parallel to the datum line or surface, as shown in Fig. 6.49(b).

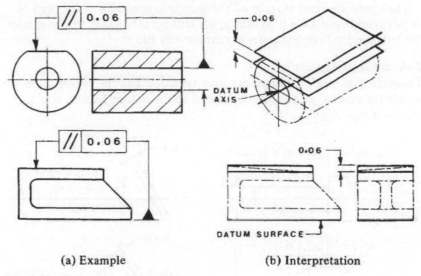

(a) Example (b) Interpretation

Fig. 6.49 Tolerances of parallelism

Tolerances of squareness

Theoretical or perfect squareness may be defined as the condition in which the feature controlled is truly perpendicular to the datum feature.

73

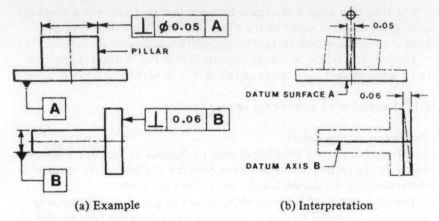

(a) Example   (b) Interpretation

Fig. 6.50 Tolerances of squareness

The axis of the vertical pillar in Fig. 6.50(a) is required to be contained within a tolerance cylinder of 0.05 mm diameter, the axis of which is perpendicular to the datum surface A, as shown in Fig. 6.50(b).

Note that the tolerance value is here preceded by the symbol Ø.

The controlled end surface of the second component is required to lie between two planes 0.06 mm apart and perpendicular to the axis of the left-hand cylindrical portion (datum axis B).

Tolerances of angularity

Theoretical or perfect angularity may be defined as the condition in which the controlled feature is inclined to the datum feature at a specified true angle.

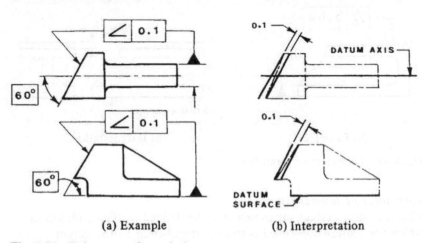

(a) Example   (b) Interpretation

Fig. 6.51 Tolerances of angularity

The tolerance zone for controlling errors of angularity is the area between two parallel straight lines or the space between two parallel planes which are inclined to the datum feature at a specified angle. The tolerance value is the distance separating the lines or planes.

The controlled inclined surfaces of the parts in Fig. 6.51(a) are to lie between two planes 0.1 mm apart which are inclined at 60° to the datum axis of the cylindrical portion or the datum surface as shown in Fig. 6.51(b).

6.18 Tolerances of location for related features

Tolerances of position

The theoretical position of a feature is the specified true position of the feature as located by boxed dimensions.

The actual point shown in Fig. 6.52(a) is required to lie within a tolerance circle 0.1 mm diameter centred on the specified true point of intersection, as shown in Fig. 6.52(b).

The axis of the hole is required to be contained within a tolerance cylinder 0.08 mm diameter centred on the specified true position of the axis of the hole.

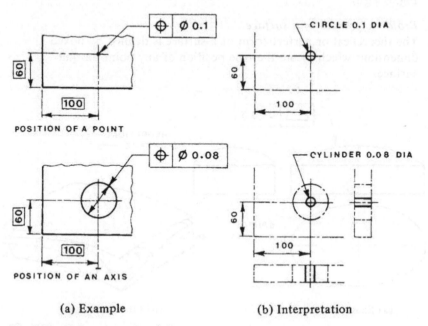

(a) Example   (b) Interpretation

Fig. 6.52 Tolerances of position

Tolerances of concentricity

Theoretical or perfect concentricity may be defined as the condition in which the controlled features (which may be circles or cylinders) lie truly on the same centre or axis as the datum features.

The tolerance zone for controlling errors of concentricity is a circle or cylinder within which the centre or axis of the controlled feature is to be contained. The tolerance value is the diameter of the tolerance zone.

The axis of the right-hand cylindrical portion of the component in Fig. 6.53(a) is to be contained within a cylinder 0.08 mm diameter and is to be co-axial with the axis of the left-hand portion, which is the datum, as shown in Fig. 6.53(b).

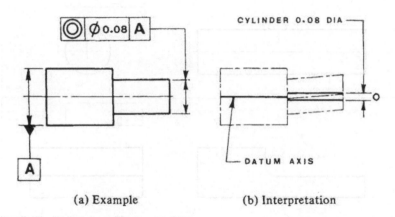

(a) Example   (b) Interpretation

Fig. 6.53 Tolerance of concentricity

Tolerances of symmetry

Theoretical or perfect symmetry may be defined as the condition in which the position of the feature is specified by its perfect symmetrical relationship to a datum.

The tolerance zone for controlling errors of symmetry is the area between two parallel lines or the space between two parallel planes which are symmetrically disposed about the datum feature.

The median plane of the slot controlled in Fig. 6.54(a) is required to lie between two parallel planes 0.08 mm apart which are symmetrically disposed about the datum plane, as shown in Fig. 6.54(b).

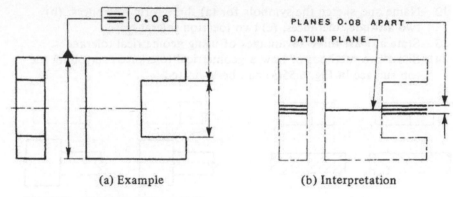

(a) Example   (b) Interpretation

Fig. 6.54 Tolerance of symmetry

6.19 Test questions

1 Define a tolerance.
2 Show by means of neat sketches three possible errors when producing a hole in a component.
3 Name and sketch three main types of engineering fit.
4 Calculate the maximum and minimum limits for the following shaft and hole nominal sizes: (a) 50 H11/c11, (b) 100 H7/n6, (c) 150 H7/s6.
5 With the help of simple sketches, show how limits between the centres of holes can be indicated by (a) chain dimensioning, (b) parallel dimensioning. Identify the main disadvantage of chain dimensioning.
6 The limits between centres of four holes A, B, C, and D are indicated by chain dimensioning in mm:

limits between holes A and B are	120.02
	119.98
limits between holes C and D are	130.02
	129.98
limits between holes A and D are	360.00
	359.88

Calculate the limits between holes B and C if all holes lie in sequence along the same centre line.

7 Define a geometrical tolerance.
8 State at least three reasons for using geometrical tolerances.
9 Define (a) a tolerance zone, (b) a datum feature.
10 Show how geometrical tolerances can be indicated by means of a rectangular frame.
11 Indicate the difference between form and attitude tolerances.

12 Name and sketch the symbols for (a) three form tolerances, (b) two attitude tolerances, (c) two location tolerances.

13 State at least three advantages of using geometrical tolerances.

14 Show two methods of how a geometrical tolerance controlling the top surface in Fig. 6.55(a) can be indicated.

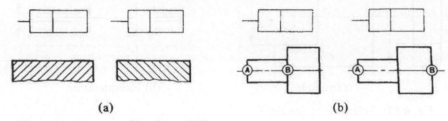

(a)   (b)

Fig. 6.55 Test questions 14 and 15

15 If the geometrical tolerance refers to the axis between A and B, complete Fig. 6.55(b) and indicate the position of the arrowheads.

16 Show two methods of how the top surface of the component in Fig. 6.56(a) can be indicated as the datum.

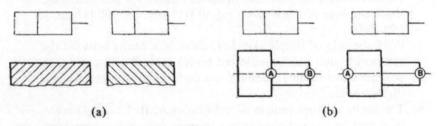

(a)   (b)

Fig. 6.56 Test questions 16 and 17

17 If the axis between A and B is the required datum feature, show how this can be indicated in Fig. 6.56(b).

18 Complete the drawings in Fig. 6.57 by indicating the correct geometrical tolerances to satisfy the following conditions in each case: (a) the axis of the whole component is required to be contained in a cylindrical zone 0.03 mm diameter; (b) the top surface of the component is required to lie between two parallel planes 0.03 mm apart; (c) the periphery at any cross-section perpendicular to the axis is required to lie between two concentric circles 0.03 mm radially apart; (d) the right-hand face of the component is required to lie between two parallel planes 0.03 mm apart and perpendicular to the top surface.

Tracing paper may be used.

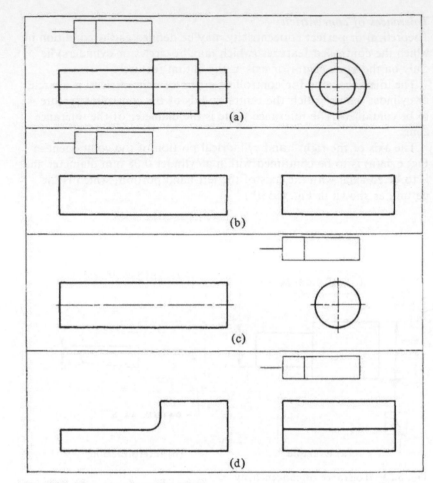

Fig. 6.57 Test questions 18 and 19

19 Complete the drawings in Fig. 6.57 by indicating the correct geometrical tolerances to satisfy the following conditions in each case: (a) the axes of the right-hand and left-hand cylindrical portions are required to be contained within one cylinder 0·02 mm diameter; (b) the top face of the part is required to lie between two parallel planes 0.08 mm apart which are perpendicular to the datum plane, which is the right-hand face; (c) the curved surface of the part is required to lie between two cylindrical surfaces co-axial with each other, a radial distance of 0.03 mm apart; (d) the top surface of the part is required to lie between two planes 0.05 mm apart and parallel to the datum plane, which is the bottom plane.

Tracing paper may be used.

20 By means of neat sketches and explanatory notes, interpret the geometrical tolerances in Fig. 6.58. Tracing paper may be used.

21 Name and interpret by means of neat sketches and explanatory notes the geometrical tolerances shown in Fig. 6.59. Tracing paper may be used.

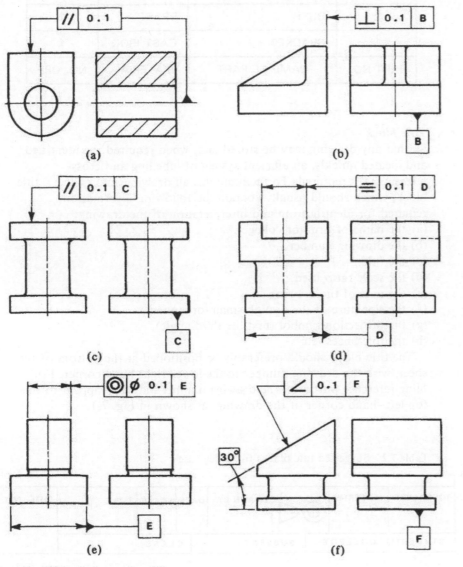

Fig. 6.58 Test question 20

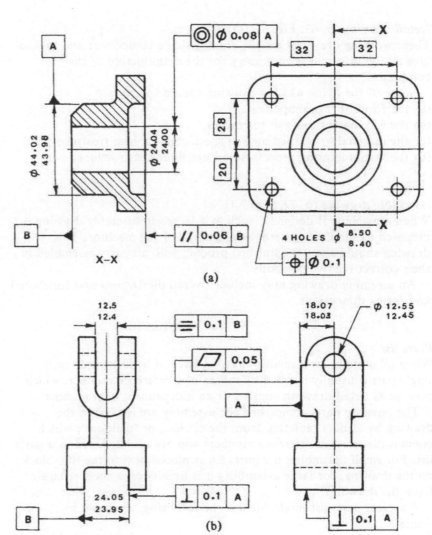

Fig. 6.59 Test question 21

7 Detail and assembly drawings

Most of the problems in this chapter require the solutions in the form of two main types of engineering drawing: detail and assembly.

Detail drawing (p. 63, Fig. 6.21(a))

These working drawings usually show a single component and should give all the information necessary for the manufacture of the component:

Some of the items which a drawing should specify are
(a) the form of the component,
(b) the full dimensions and tolerances,
(c) the material to be used and its specifications, heat treatment, etc,
(d) the manufacturing processes and machining instructions.

Assembly drawing (p. 79, Fig. 7.1)

When a machine is designed, such as a lathe, an assembly drawing is prepared to show the general arrangement of the machine. This drawing should show the finished product with all parts assembled in their correct relative positions.

An assembly drawing may include overall dimensions and functional and fitting dimensions.

Parts list

When all parts in an assembly drawing have to be identified, each single part is usually labelled by means of a reference number, which may be its detail-drawing number or an independent item number.

The separate parts comprising the assembly are located in the drawing by leaders radiating from the circles , or 'balloons', which contain the relevant reference numbers and are usually listed in a parts list. For small assemblies the parts list is placed next to the title block on the drawing, for large assemblies it is usually on a sheet separate from the drawing.

A typical parts list might include the following, as shown in Table 7.1:
(a) the part number,
(b) the name or description of the part,
(c) the material from which the part is to be made,
(d) the quantity required.

Table 7.1 Parts list

3	SMALL PULLEY	NYLON	5
2	BUSH	BRASS	2
1	BRACKET	CAST IRON	1
PART NO.	NAME OF PART	MATERIAL	NO. OFF

Title block

So that any drawing may be stored and, when required, be identified and located quickly, an efficient system of labelling and cross-referencing is required. To facilitate this all drawings must have a title block, which should usually contain the following information required for identification and interpretation of the drawing:
(a) the name of firm (or college),
(b) the drawing number,
(c) the title,
(d) the scale ratio used,
(e) the date of the drawing,
(f) the signature of the draughtsman (or student),
(g) the projection symbol (first- or third-angle),
(h) the tolerances, etc.

The title block should preferably be positioned at the bottom of the sheet, with the drawing number in the lower right-hand corner. For filing reference purposes, the drawing number may also appear in the top left—hand corner of the drawing, as shown in Fig. 7.1.

Table 7.2 Suggested title blocks for college use

TOLERANCE	MATERIAL	⊕◁	DRAWN BY	DATE	SCALE	TITLE	DRG.NO.
STUDENT'S COLLEGE	SUBJECT				CLASS		

STUDENT'S COLLEGE	DRAWN BY	DATE	SCALE 1:1	◁⊕	TITLE	DRG NO.

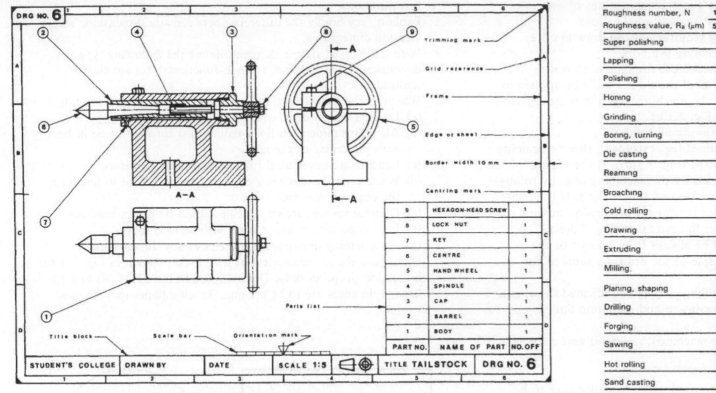

Fig. 7.1 Assembly drawing with explanatory notes

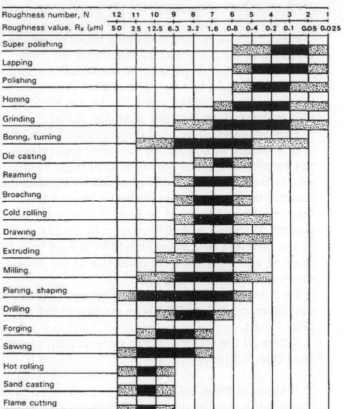

Fig. 7.2 Surface roughness values produced by common production processes

Assembly drawings may include a grid reference system which is based on numbered and lettered divisions in the margin of the drawing sheet (Fig. 7.1).

Central marks may be used for the positioning of drawing for reproduction processes and a figureless scale bar to indicate the proportional scale.

The orientation mark may be used to indicate the format of the drawing: *Landscape* (Fig. 7.1) with the horizontal sides being longest or *portrait* (p. 83, Fig. 7.9), with the vertical sides being longest. It is positioned at the mid-point of the lowest part of the drawing.

The minimum width of the border should be 10 mm.

General notes should preferably be grouped together and specific notes should appear near to their relevant features. The minimum height of characters should be 2.5 mm for dimensions and notes and 7 mm for drawing numbers and titles.

7.1 Surface texture

In general engineering, all manufactured surfaces depart to a larger or lesser extent from an ideally smooth surface.

The usual surface texture, when magnified many times, is represented by irregularities forming a series of peaks and valleys varying in height and spacing. This departure of the surface profile below and above the centre line representing the ideal smooth surface, is measured in micrometres, thousandths of a millimetre, as an arithmetical mean deviation. The centre line is constructed in such a way that within the predetermined sampling length the sum of all the areas between the centre line and the surface profile on each side of it are equal, as shown in Fig. 7.3(a).

The nominal surface texture numbers, R_a, are shown in Fig. 7.2 with the ranges of roughness values for typical common production processes. Dark areas represent the average applications of processes whereas shaded areas represent less frequent applications.

Clear instructions for machining requirements on drawings are indicated by British Standard machining symbols.

To indicate the surfaces to be machined, the symbol shown in Fig. 7.3(b) should be used. The symbol should normally be applied to the line representing the surface to be machined, or it may be applied to a leader or extension line as in Fig. 7.3(h).

Thin lines should be used for drawing all symbols.

If a surface should not be machined (eg. extruded), then 'machining not permitted' symbol should be used (Fig. 7.3(d)). If the maximum roughness can be obtained not necessary by machining (e.g. by rolling) 'machining optional' symbol should be used (Fig. 7.3(c)). If the final texture is to be produced by surface treatment, production method or coating, it should be indicated over the extension (Fig. 7.3(e)).

The direction of reading should be always from bottom of the drawing and from the right-hand side of the drawing, same as for dimensioning (Fig. 7.3(i)).

Permissible values of surface roughness may be indicated showing a maximum value (Fig. 7.3(f)) or maximum and minimum limits (Fig. 7.3(g))

Where all the surfaces are to be machined, a general note may be used (Fig. 7.3(j)).

Smooth surfaces are expensive to produce and usually have to be inspected, hence the designers should always keep unnecessary surface requirements to a minimum.

7.2 Test questions

1 Explain very briefly the difference between size dimensions and location dimensions.
2 With the help of simple sketches, define the following types of dimension: (a) functional, (b) non-functional, (c) auxiliary (redundant).
3 Which of the following statements are true and which are false?
  (a) Leaders end in dots, when crossing the outline.
  (b) All radius dimension lines should pass through or be in line with the centre of the arc.
  (c) Leaders are never used for dimensioning diameters.
  (d) When a leader touches a circle, it should not be in line with the centre of the arc.
  (e) Dimensions not drawn to scale should be put in brackets.
  (f) Inspection dimensions always include tolerances.
  (g) An assembly drawing never includes any dimensions.
4 Dimension the sectional view of the bracket shown in Fig. 7.4 for inspection purposes only. The holes are to have H11 fits and the linear tolerances are to be 0.1 mm. Tracing paper may be used.

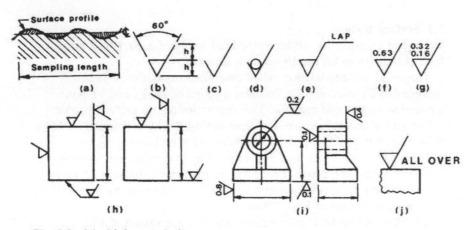

Fig. 7.3 Machining symbols

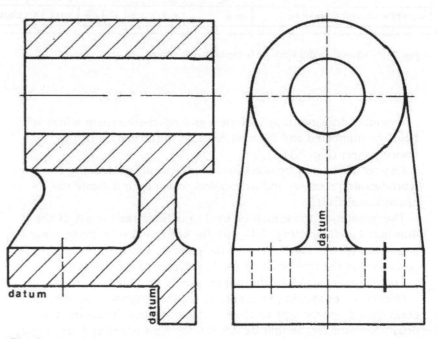

Fig. 7.4 Test questions 4 and 5

80

5 Add a plan view and dimension the bracket shown in Fig. 7.4 for casting, including all necessary instructions. Tracing paper may be used.

6 The bracket shown in Fig. 7.5 is drawn to half full size (1:2).

Redraw the bracket full size and dimension it for casting, including all necessary instructions. (Alternatively, tracing paper may be used for a half-full-size drawing).

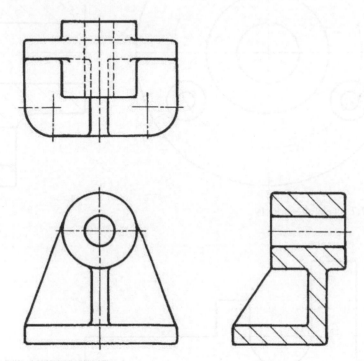

Fig. 7.5 Test question 6

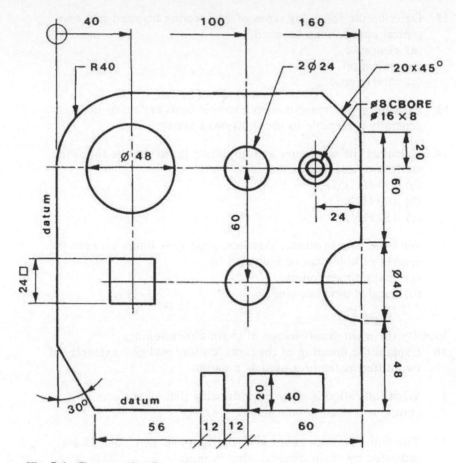

Fig. 7.6 Test question 7

7 (a) Figure 7.6 shows a fully dimensioned template. Assume all numerical values to be in mm and to be correct. How many dimensions are incorrectly given with respect to BS 308 – 10? 11? 12? 13? 14? 15? or 16?

   (b) Place a tracing paper over Fig. 7.6 and trace the given view. Add a plan assuming the thickness of the template to be 20 mm. Redimension the template correctly utilising the datum surfaces shown.

   (c) Redraw the template full size and add a plan assuming the thickness of the template to be 20 mm. Redimension the template using a H8 fit for all holes.

8 Define the term 'tolerance'

9 Explain why tolerances are used for dimensioning purposes.

10 Discuss briefly the following considerations in establishing reasonable tolerances:

   (a) function,
   (b) interchangeability,
   (c) cost.

11 Define the following terms as specified by BS 4500:

   (a) basic size,
   (b) deviation,
   (c) maximum limit of size.

12 Describe the following types of engineering fits, and give two typical applications for each:
   (a) clearance,
   (b) transition,
   (c) interference.

13 Explain, giving reasons, why the hole-basis system of fits is generally preferable to the shaft-basis system.

14 Calculate the maximum and minimum limits for the following shaft and hole nominal sizes:
   (a) 50 H11/c11,
   (b) 100 H7/n6,
   (c) 150 H7/s6.

15 With the help of simple sketches, show how limits between the centres of holes can be indicated by
   (a) chain dimensioning,
   (b) parallel dimensioning.

Identify the main disadvantage of chain dimensioning.

16 Explain the meaning of the term 'datum' and give examples of two different features used as a datum.

17 What indication is given on a drawing that all surfaces of a component must be machined?

18 The limits between centres of four holes A, B, C, and D are indicated by chain dimensioning in mm:

   limits between holes A and B are   20.02
                                       19.98
   limits between holes C and D are   30.02
                                       29.98
   limits between holes A and D are   60.00
                                       59.88

Calculate the limits between holes B and C if all holes lie in sequence along the same centre line.

19 Fully dimension the component shown in Fig. 7.7. Choose your own datum, include machining symbols and the cutting plane, and indicate the angle of projection.

20 Fully dimension the component shown in Fig. 7.8. Tracing paper may be used.

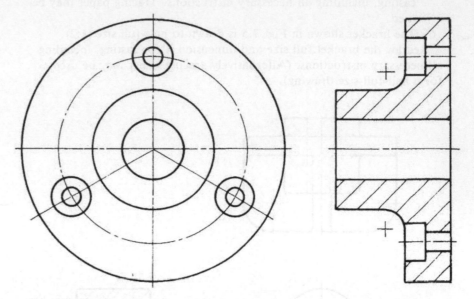

Fig. 7.7 Test question 19

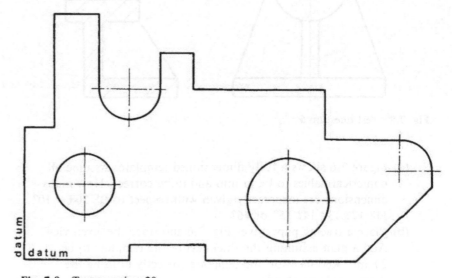

Fig. 7.8 Test question 20

82

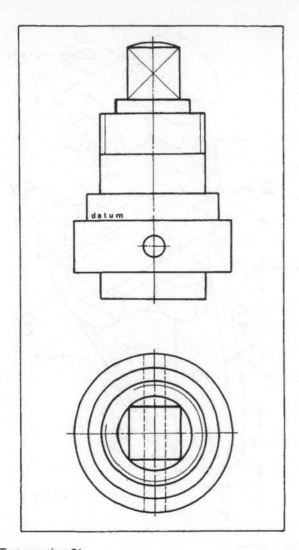

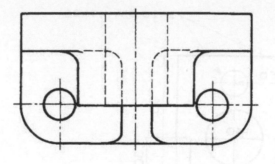

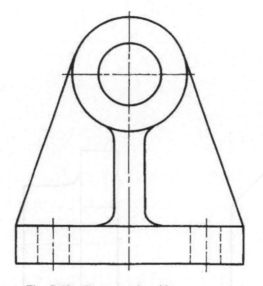

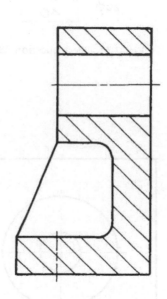

Fig. 7.10   Test question 22

Fig. 7.9   Test question 21

21   Fully dimension the component shown in Fig. 7.9 and include the projection symbol. Tracing paper may be used.

22   The bracket shown in Fig. 7.10 is drawn to half full size (1:2).
Redraw the bracket full size, and dimension it for machining processes. Include all required machining symbols, and tolerance the horizontal hole using a H9 fit for a 32 mm diameter nominal size. (Alternatively, tracing paper may be used for a half-full-size drawing.)

23   Figure 7.11 shows a partially dimensioned template. Assume all numerical values to be in mm and to be correct.
How many dimensions are correctly given with respect to BS 308 − 0? 1? 2? 3? 4? 5? or 6?

24   Fully dimension the component shown in Fig. 7.12. Choose your own datum, include machining symbols, and indicate three functional dimensions by using a letter F.

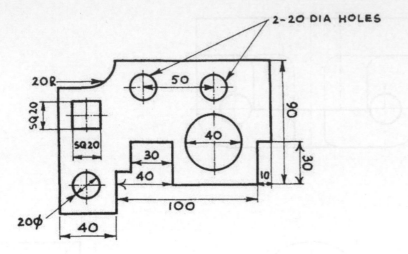

Fig. 7.11   Test question 23

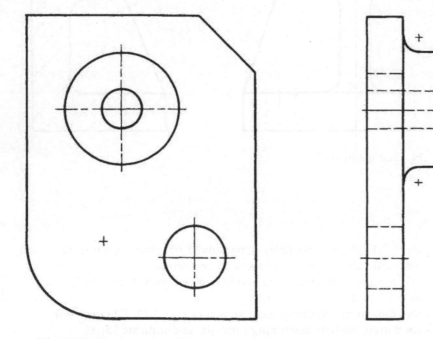

Fig. 7.12   Test question 24

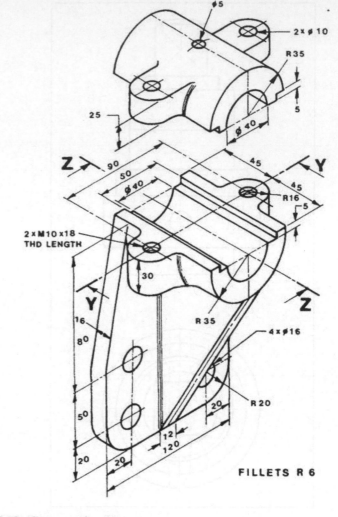

Fig. 7.13   Test question 25

25   Draw the following views of the assembled parts shown in
Fig. 7.13 adding studs, nuts, washers, and 120 mm length of
40 mm diameter shaft:
(a) a sectional front view on YY,
(b) a sectional end view on ZZ,
(c) a plan view showing hidden detail.
Include the functional dimensions, insert a title block, and add a
parts list.

84

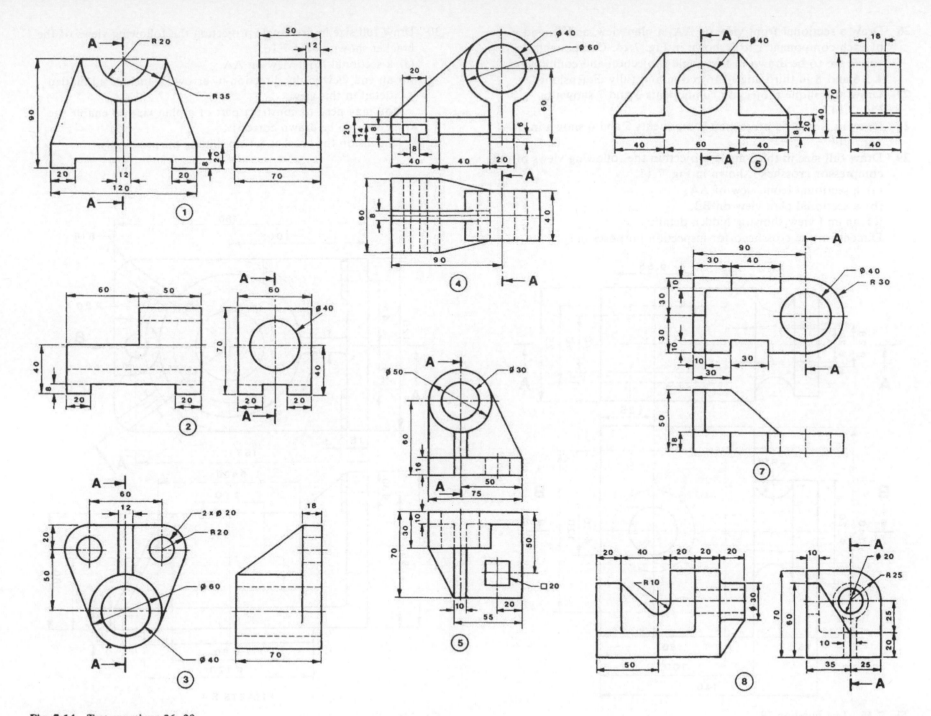

Fig. 7.14 Test questions 26–28

85

26 Draw a sectional front view on AA, a plan view, and an end view of each component 1 to 8 shown in Fig. 7.14. Components 1,2,3 and 7 are to be drawn in first-angle projection and components 4.5.6 and 8 in third-angle projection and fully dimensioned.

27 Draw in oblique projection components 3 and 7 shown in Fig. 7.14.

28 Draw in isometric projection components 5 and 6 shown in Fig. 7.14.

29 Draw full size in third-angle projection the following views of the compressor crosshead shown in Fig. 7.15.
(a) a sectional front view of AA,
(b) a sectional plan view on BB,
(c) an end view showing hidden detail.
Dimension the crosshead for inspection purposes only.

30 Draw full size in first-angle projection the following views of the bracket shown in Fig 7.16.
(a) a sectional front view on AA.
(b) an end view in the direction of arrow B. Include all hidden detail in this view.

You may need to construct part of a plan view to enable the front view to be drawn correctly.

Dimension the bracket for machining purposes only.

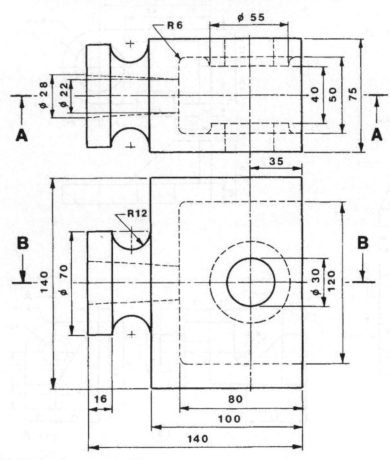

Fig. 7.15   Test question 29

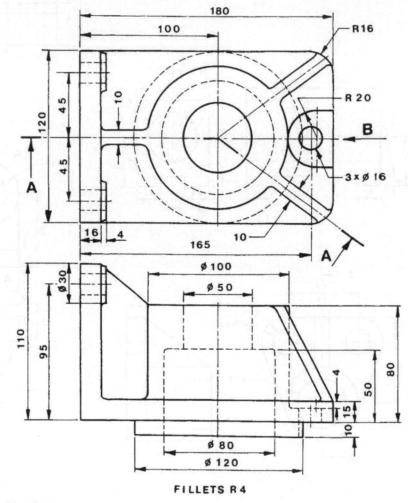

FILLETS R 4

Fig. 7.16   Test question 30

86

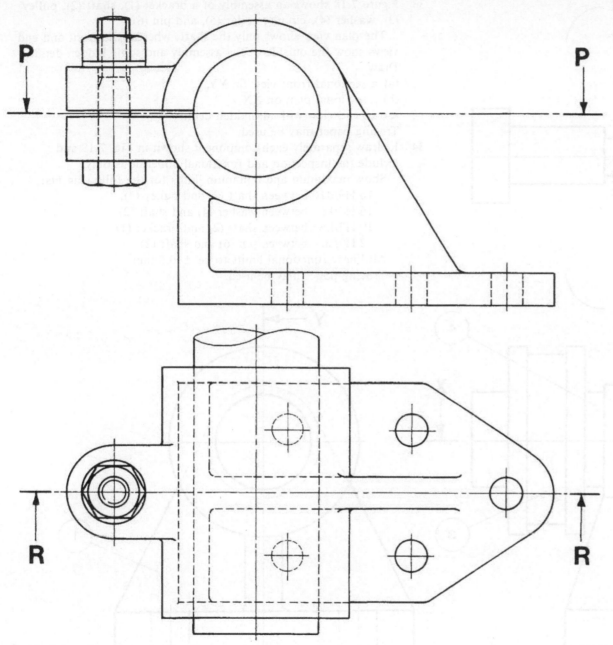

Fig. 7.17 Test questions 31 and 32

31 The clamp unit shown in Fig. 7.17 consists of a bracket (item 1), a shaft (item 2), a bolt (item 3), a nut (item 4) and a washer (item 5).

Place a tracing paper over the figure and draw:

(a) a sectional front view on RR,

(b) a sectional plan view on PP.

Balloon reference the assembly, complete the parts list, insert the projection symbol and print the title *clamp unit*.

32 The clamp unit shown in Fig. 7.17 consists of a bracket (item 1), a shaft (item 2), a bolt (item 3), a nut (item 4) and a washer (item 5).

Draw full size with all parts assembled in third-angle projection,

(a) a sectional front view on RR,

(b) a sectional plan view on PP,

(c) an end view with hidden detail.

Balloon reference the assembly, complete the parts list, insert the projection symbol and a title block.

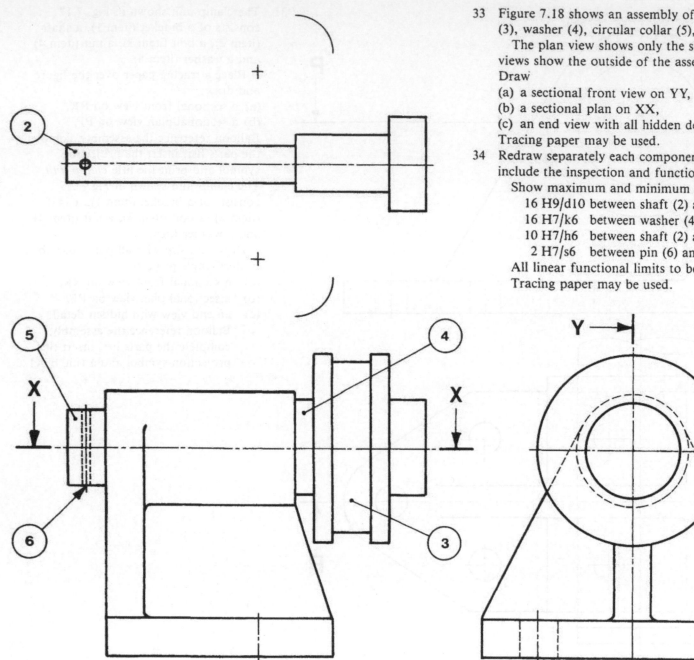

33 Figure 7.18 shows an assembly of a bracket (1), shaft (2), pulley (3), washer (4), circular collar (5), and pin (6).

The plan view shows only the shaft, whereas the front and end views show the outside of the assembly and some hidden detail. Draw

(a) a sectional front view on YY,

(b) a sectional plan on XX,

(c) an end view with all hidden detail.

Tracing paper may be used.

34 Redraw separately each component shown in Fig. 7.18 and include the inspection and functional dimensions only.

Show maximum and minimum limits for the following fits:

16 H9/d10 between shaft (2) and pulley (3),

16 H7/k6 between washer (4) and shaft (2),

10 H7/h6 between shaft (2) and bracket (1),

 2 H7/s6 between pin (6) and shaft (2).

All linear functional limits to be ± 0.5 mm.

Tracing paper may be used.

Fig. 7.18 Test questions 33 and 34

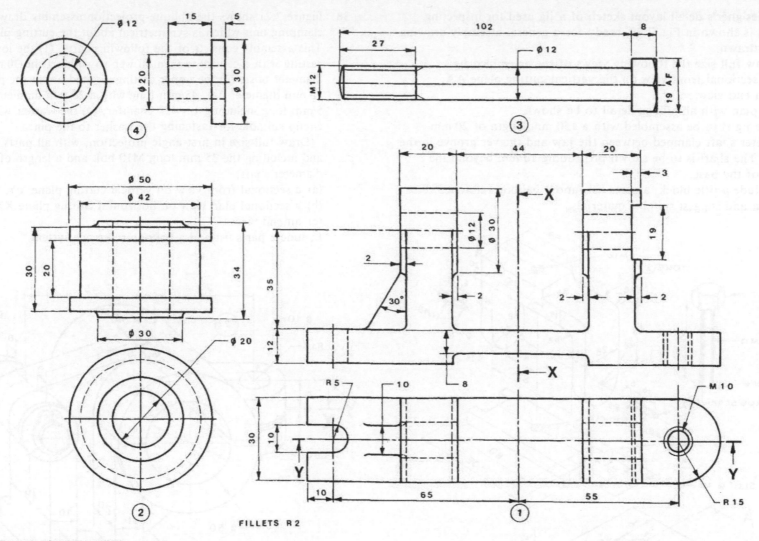

Fig. 7.19 Test questions 35 and 36

35 The belt-pulley unit shown in Fig. 7.19 consists of a belt pulley (2), a mounting bracket (1), a fitted bolt (3), and two bushes (4).

Draw full size with all parts assembled, including a suitable nut with locking device (see Chapter 10),

(a) a sectional front view on YY,

(b) a sectional end view on XX,

(c) a plan view. Show all hidden detail in this view.

Insert a title block and add a parts list.

36 Dimension items (2), (3) and (4) in Fig. 7.19 for inspection purposes only, with the following fits:

H7/p6 between shaft (3) and two bushes (4),

H9/e9 between two bushes (4) and belt pulley (2).

Linear tolerances to be ± 0.2 mm.

37 The designer's detail layout sketch of a jig used for inspecting shafts is shown in Fig. 7.20, ready for a general-assembly drawing to be drawn.

Draw full size the following views of the assembled jig:
(a) a sectional front view on the vertical cutting plane AA,
(b) an end view,
(c) a plan with all hidden detail to be shown.

The jig is to be assembled with a 150 mm length of 20 mm diameter shaft clamped between the jaw and the vee groove in the base. The shaft is to be shown protruding 10 mm beyond the right of the base.

Include a title block, a parts list, and a balloon reference system and suggest suitable materials.

38 Figure 7.21 shows the oblique-projection assembly drawing of a clamping unit which is symmetrical about the cutting plane YY. This assembly consists of the following parts: (1) the lower casting with a central tangential web to support the 30 mm diameter boss; (2) the upper casting pivoted about the pin; (3) the 10 mm diameter pin, 45 mm long with a 20 mm diameter head, 5 mm long including 1 × 45° chamfer; (4) the washer with a circlip suitable for fastening the washer to the pin.

Draw full size in first-angle projection, with all parts assembled and including the 35 mm long M10 bolt and a length of 30 mm diameter shaft,
(a) a sectional front view on vertical cutting plane YY,
(b) a sectional plan view on horizontal cutting plane XX,
(c) an end view.
Include a parts list and a balloon reference system.

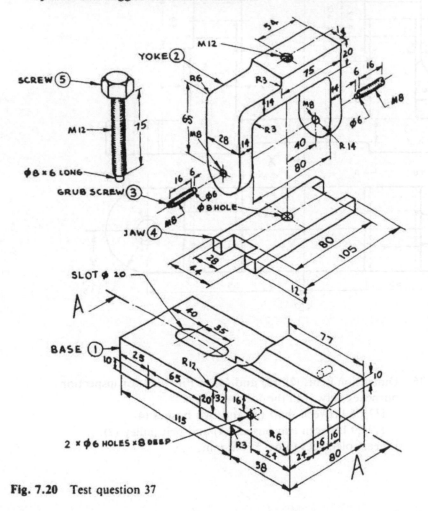

Fig. 7.20 Test question 37

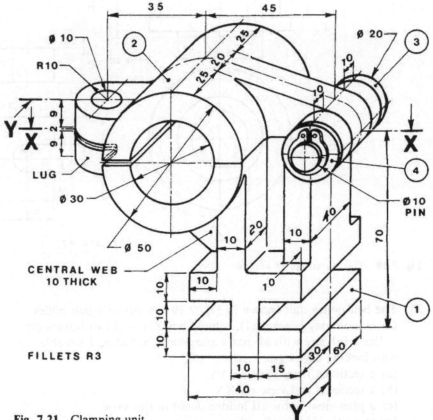

Fig. 7.21 Clamping unit

90

39 The adjusting unit shown in Fig. 7.22 consists of a body (1), a handle (2), a pivot (3), a screw (4), a collar (5), and a pin (6).

Draw the following views of the completely assembled adjusting unit:

(a) a front view in the direction of arrow B, including all hidden detail;

(b) a sectional end view on the cutting plane AA.

Include the main functional dimensions that have a bearing on the assembly. Insert a title block, with the relevant information, and a parts list with the corresponding reference balloons and suggest suitable materials.

40 Figure 7.22 shows the adjusting unit of an instrument used for electrical adjusting purposes. By moving the handle (2), the required position of the slot in the pivot (3) is obtained.

Draw twice full size the pivot, incorporating the following conditions:

(a) a fit of 8 H7/h6 is required between the pivot and the body (1), and a fit of 10 H7/p6 between the pivot and the handle (2);

(b) the axis of the pivot is required to lie between two parallel planes 0.06 mm apart which are symmetrically disposed about the median plane of the slot in the end of the pivot.

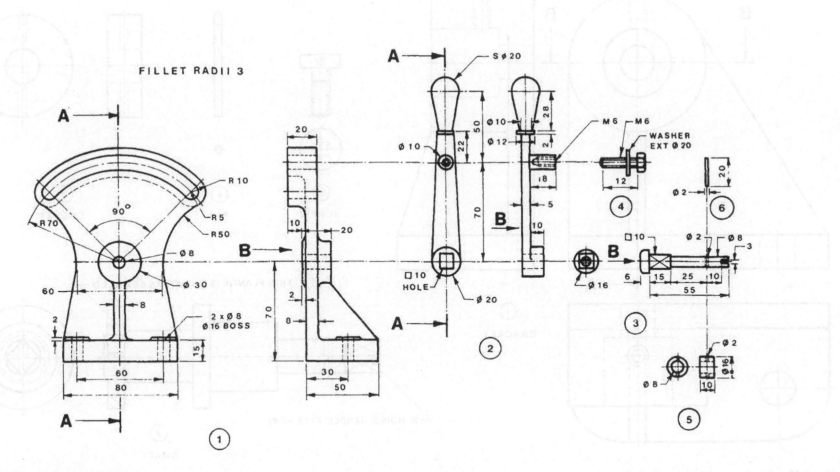

Fig. 7.22 Electricity adjusting unit

91

41 Figure 7.23 shows the components, drawn full size, of an electrical *hand coil winder*.

The coil former, not shown, is fitted between the shaft flange (2) and the flange (6). The securing screw (5) is used to lock the winder after the process of winding the coil is completed.

To simplify the assembly all undercuts are not shown and all bearings have been omitted.

Draw, excluding the former, in third-angle projection with all parts assembled:
(a) a front view,
(b) a sectional end view on AA,
(c) a sectional plan view on BB of part (1) i.e. bracket only.

To complete the assembly, design a winding handle (7) which is to fit on the square end of the shaft (2). This may be shown in view (b) only.

Insert a title block, a parts list and a reference system.

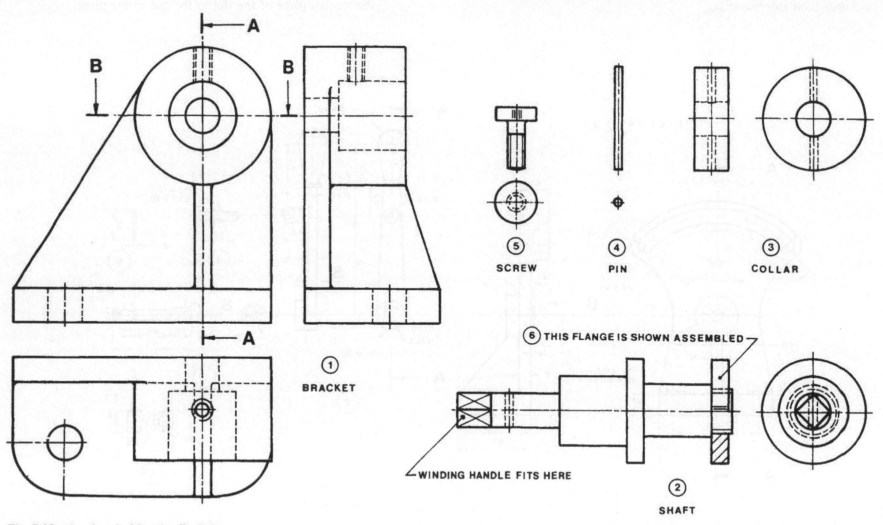

Fig. 7.23 An electrical hand coil winder

42 The flange marked 'A' in Fig. 7.24 is to accommodate a 10 mm × 10 mm square key and a 40 mm diameter 150 mm long shaft.

Select a scale and draw (a) the flange, (b) the square key, and (c) the shaft, and then tolerance them according to the following fits: H7/h6 fit between the flange and the shaft; H7/k6 fit between the key and the flange and between the key and the shaft.

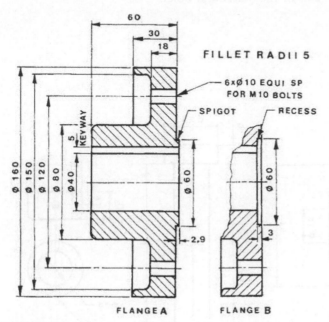

Fig. 7.24 Flange coupling

43 A shaft coupling connecting two co-axial shafts consists of two flanges, identical except for spigot and recess shown in Fig. 7.24: two keys; two shafts; and six bolts, nuts, and washers.

(a) Select a scale and draw a half-sectional front view of the assembled coupling, showing two shafts in position. The required view should show the assembled coupling in section above the shaft centre line and it must include one of the bolts.

The remaining part of the front view, below the shaft centre line, is to show the outside view of the assembled coupling.

No hidden detail is required.

(b) Draw an end view including all hidden detail.

Add a parts list with a corresponding balloon-reference system and suggest suitable materials.

44 Draw the following views of the bearing-and-bracket assembly shown in Fig. 7.25:

(a) a sectional front view, viewed in the direction of arrow B, through the centre of the vertical web;

(b) a half-sectional end view on AA, the half-section to be to the right of the vertical centre line of symmetry.

Include one M12 nut, washer, and bolt.

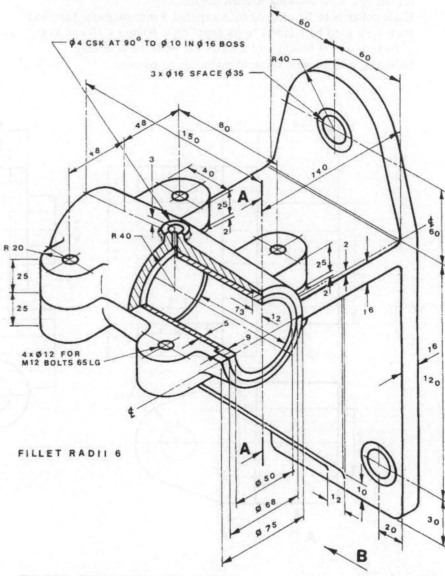

Fig. 7.25 Bearing-and-bracket assembly

45 Figure 7.26 shows a Hooke's coupling, which is used to connect two shafts that have a small degree of misalignment. It consists of two forks (1), one centre (2), two pins (3), and two collars (4).

Select a scale and draw the following views with all parts correctly assembled:
(a) a sectional front view in the direction of arrow A,
(b) a plan view,
(c) an end view showing hidden detail.

Each collar is to be secured by a tapered 3 mm diameter pin, and each fork is to be secured to its shaft by a 10 mm × 10 mm key.

Include a title block, add a parts list with the reference balloons, and suggest suitable materials to be used.

46 Figure 7.26 shows a Hooke's coupling. The centre (2) basically consists of two tubular pieces cast together at 90° to each other.

Select a scale and draw the centre and fully dimension it, satisfying the following conditions:
(a) the fit between the centre and the pins (3) is to be H7/n6,
(b) the axis of the vertical hole is required to lie between two parallel planes 0.04 mm apart which are perpendicular to the axis of the horizontal hole.

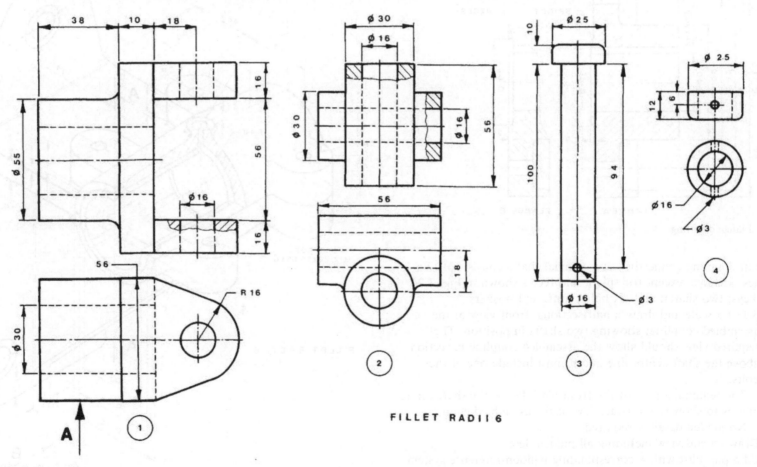

FILLET RADII 6

Fig. 7.26 Hooke's coupling

47 Figure 7.27 shows five detail parts, which together make up a
toolmaker's vice. This consists of a body (1), a moving jaw (2), a
bottom plate (3), a screw (4), a special grub screw (5), and two
M4 countersunk screws 16 mm long, which are not shown.

Select a scale and draw in correct projection the following
views of the fully assembled toolmaker's vice:
(a) a sectional front view, taking the screw axis to be a cutting
plane;
(b) an end view;
(c) a plan.
Show all hidden detail in views (b) and (c).

Include a title block with all relevant information and add a
parts list with a balloon-reference system.

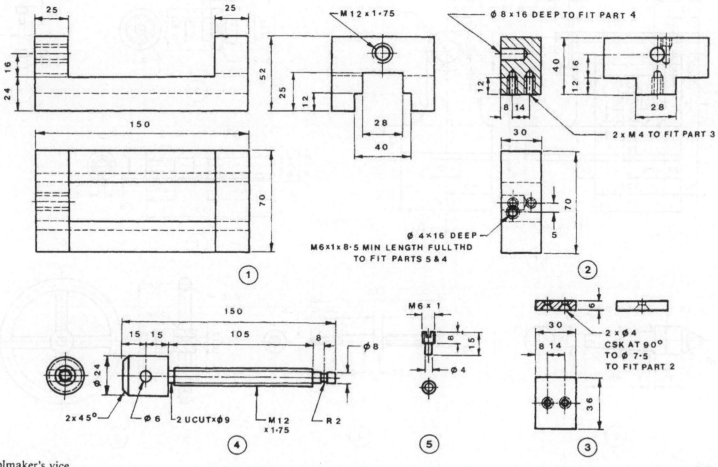

Fig. 7.27 Toolmaker's vice

48 The component parts of a tailstock are shown in Fig. 7.28. They consist of a body (1); a barrel (2); a cap (3); a spindle (4); a handwheel (5); a centre (6); a key (7); an M10 hexagon-head screw, 50 mm long with 20 mm minimum length of full thread to fit part (1); and an M10 nut, 8 mm high with a 2 mm thick washer to fit part (4).

With all parts correctly assembled, draw full size the following views:

(a) a sectional front view on AA,

(b) an end view,

(c) a plan.

Include a title block with all relevant information and reference balloons with a parts list, suggesting suitable materials.

For solution see p. 79, Fig. 7.1.

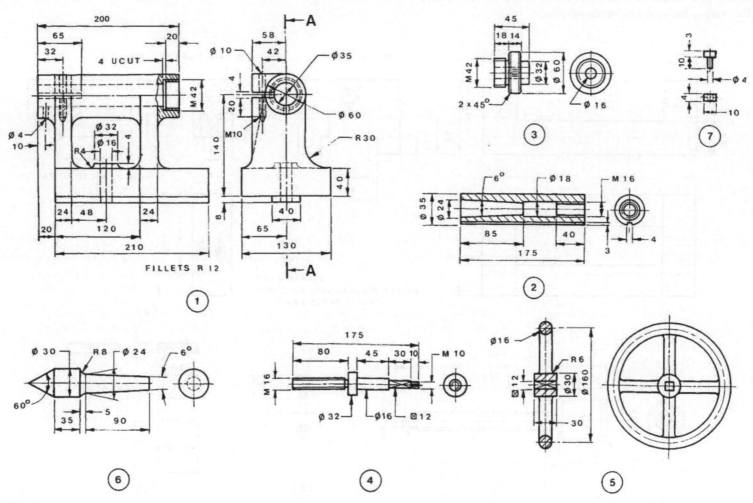

Fig. 7.28 Tailstock

96

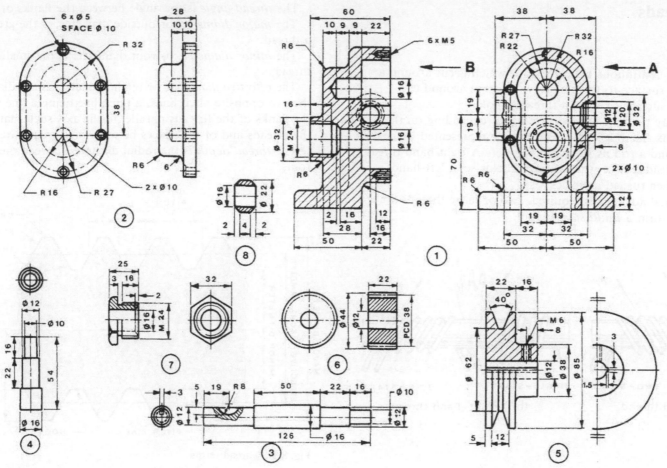

Fig. 7.29 Oil gear pump

49 Figure 7.29 shows the detail parts of an oil gear pump. The rotation of the two gears (item 6) mounted on the spindle (item 4) and the driving spindle (item 3) inside the body (item 1) and cover (item 2) allows the required flow of oil. For assembling the body and cover the M5 socket screws are to be used. Outside the pump the driving pulley (item 5) is to be fixed to the driving spindle (item 3) by means of the 6 mm radius woodruff key (item 9 not shown) and the M6 grub screw (item 10 not shown). To ensure that there is no leakage of oil the gland (item 7) with the felt (item 8) are inserted between the driving spindle (item 3) and the

M24 hole in the body (item 1). No pipes or connections are required.

With all mentioned parts assembled, draw in the first-angle projection:

(a) a sectional front view in the direction of arrow A,

(b) a plan,

(c) an end view in the direction of arrow B with the cover (item 2) and M5 screws removed.

Include a title block and a balloon reference system with a parts list.

97

8 Screw threads

A screw thread is a continuous helical groove which is cut around a cylindrical external surface to form a screw or is cut around a cylindrical internal surface to form a threaded hole.

Screw threads may be right-hand or left-hand, depending on the direction of the helix (see page 102). This can be represented by heavy strings wound around a rod as shown in Fig. 8.1. A right-hand thread advances into a threaded hole when turned clockwise; a left-hand thread advances when turned anticlockwise.

When a quick axial advance is required, two or more threads are cut side by side to form a *multi-start* thread.

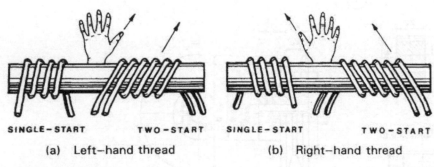

| (a)   Left–hand thread | (b)   Right–hand thread |

Fig. 8.1   Threads

8.1 Thread terms (Fig. 8.2)

The *pitch* of a thread is the distance from a point on one thread to the corresponding point on the adjacent thread, measured parallel to the axis of the thread.

The *lead* is the axial distance moved by a screw during one revolution. For a single-start thread the lead is equal to the pitch; for a two-start thread the lead is equal to twice the pitch; for a three-start thread the lead is three times the pitch; and so on (see page 104).

The *crest* of a thread is the most prominent part of an external or internal thread.

The *root* of a thread is the bottom of the groove of an external or internal thread.

The *flank* of a thread is the straight side connecting the crest to the root.

The *thread angle* is the angle between the flanks of the thread.

The *major diameter*, or outside diameter, is the greatest diameter of a thread.

The *minor diameter*, or root diameter, is the smallest diameter of a thread.

The *effective diameter*, or pitch diameter, is the diameter between the two opposite pitch lines, a pitch line being a line which intersects the flanks of the threads parallel to the axis such that the widths of the threads and of the spaces between the threads are equal.

The *thread depth* is the radial distance between the crests and the roots.

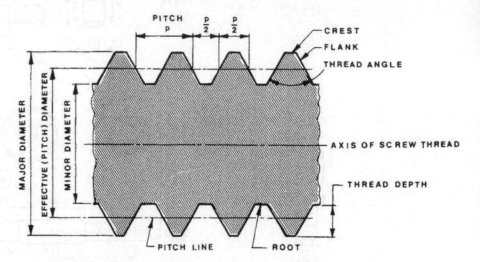

Fig. 8.2   Thread terms

8.2 Thread forms

There are two main classes of thread forms: *vee threads*, which are principally used for fastening and adjusting purposes, and *square threads*, which are used for transmitting forces (power transmission).

8.3 The ISO metric thread (Fig 8.3)

The ISO (International Organisation for Standardisation) metric thread is a vee-form thread. These are two types of ISO metric thread:

(a) *fine-pitch-series threads*, which are mainly used for special applications such as for thin-walled components, fine adjustments on machine tools, etc.;

(b) *coarse-pitch-series threads*, which are suitable for metric fasteners and for general-purpose applications.

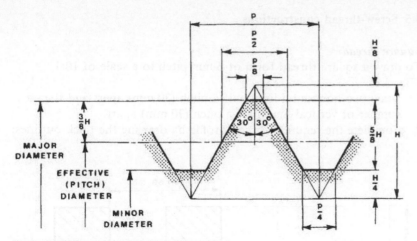

Fig. 8.3 Basic form of ISO metric thread

Classes of fit

There are three classes of fit between external (bolt) threads and internal (nut) threads for different engineering uses: close, medium, and free.

The close fit   This fine-tolerance fit is applied to threads used for very high quality precision work and requires very thorough inspection.

The medium fit   This is suitable for most general engineering purposes.

The free fit   This coarse-tolerance fit is applied where a quick and easy assembly is needed, with threads occasionally becoming dirty and slightly damaged.

Designation of ISO metric threads

The complete designation of ISO metric screw threads is shown in the following examples:

(a) For an internal thread:

   M20 × 2.5 – 6H

   └── Thread tolerance-class symbol
   └── Pitch in millimetres
   └── Nominal size (major diameter) in millimetres
   └── Symbol for ISO metric thread

(b) For an external thread:

   M20 × 2.5 – 6g

(c) For a pair of threaded parts:

   M20 × 2.5 – 6H/6g

On a drawing, the thread tolerances are usually not indicated and the ISO metric screw threads are designated as, for instance M20 × 1.5

for the fine-pitch series and M20 × 2.5 for the coarse-pitch series, or simply as M20 when the coarse threads are specified.

8.4 Other thread forms

British Standard Pipe (BSP) thread (Fig. 8.4)

This Whitworth standard-form thread has a relatively fine pitch so that both internal and external threads may be cut on thin tubing and piping. When used for specific high-pressure applications, the BSP threads are cut in a tapered rather than a cylindrical surface.

BSP threads are designated by the size of the internal diameter (bore) of the pipe, instead of by the thread major diameter as is the case for all other types of thread.

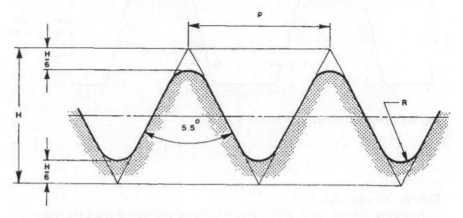

Fig. 8.4   BSP thread

Square thread (Fig. 8.5)

The square thread form is used mainly to transmit forces in both axial directions. Since its flanks are normal to its axis, it offers less frictional resistance to motion than does a vee thread. The main applications of this thread include valve spindles, machine leadscrews, vice screws, and screw-jacks.

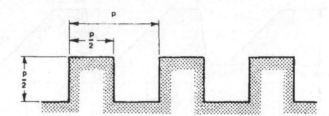

Fig. 8.5   Square thread

ISO metric trapezoidal screw thread (Fig. 8.6)

This thread is a modified form of square thread. It is stronger, because of the wider base, and easier to cut, due to its taper. The trapezoidal thread is widely used to transmit forces in valve spindles, lathe leadscrews, etc., but its inclined flanks give rise to frictional resistance. When a split clamping nut is used, as on the leadscrew of a lathe, both halves of the nut will engage easily on the tapered sides of this thread.

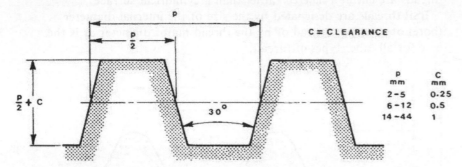

P mm	C mm
2–5	0·25
6–12	0·5
14–44	1

Fig. 8.6 ISO metric trapezoidal screw thread

Buttress thread (Fig. 8.7)

This strong thread form offers less frictional resistance than the vee-type threads and is designed to withstand heavy forces applied in one direction. Its main applications include quick-release vices, slides, and presses.

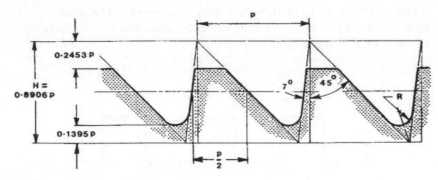

Fig. 8.7 Buttress thread

8.5 Screw-thread constructions

Square thread

To draw a square thread form of 6 mm pitch to a scale of 10:1 (Fig. 8.8):

1 Draw two horizontal lines half a pitch (30 mm) apart and then a number of vertical lines half a pitch (30 mm) apart.
2 Complete the required thread profile by drawing the thick outlines.

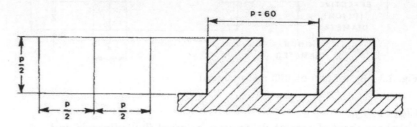

Fig. 8.8 Square thread

ISO metric trapezoidal thread

To draw an ISO metric trapezoidal thread form of 6 mm pitch to a scale of 10:1 (Fig. 8.9):

1 From a horizontal line AB representing the pitch line, draw say six flanks half a pitch (30 mm) apart and inclined at 15° to the vertical to give the thread angle of 30°.
2 On each side of the pitch line, draw a parallel line a quarter of a pitch (15 mm plus half a clearance (2.5 mm) away.
3 Complete the required thread profile by drawing the thick outlines.

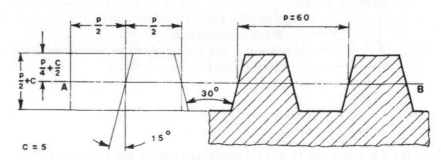

Fig. 8.9 ISO metric trapezoidal thread

Buttress thread

To draw a buttress thread of 6 mm pitch to a scale of 10:1 (Fig. 8.10):

1 From a horizontal line AB, draw say four vertical lines a pitch (60 mm) apart and then draw the flanks inclined at 7° and 45° to the vertical.
2 Join the apex points to obtain the height H of 0.8906×60 mm $= 54.44$ mm and then draw two horizontal lines, one at 0.2453×60 mm $= 14.72$ mm from AB and the other at 0.1395×60 mm $= 8.37$ mm from CD.
3 Bisect the angles marked E and F to obtain the intersection point G, and transfer this point horizontally to the corresponding positions for adjacent teeth.
4 With centres G, draw tangential arcs and complete the required thread profile by drawing the thick outlines.

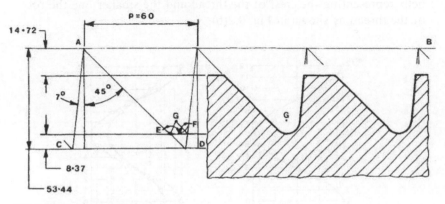

Fig. 8.10 Buttress thread

ISO metric thread

(a) To draw an *external* (bolt) thread M64 × 6 to a scale of 10:1 (Fig. 8.11);

1 From a horizontal line AB, draw say seven vertical lines half a pitch (30 mm) apart and then draw the flanks inclined at 30°.
2 Join the apex points to obtain the height H and divide this into first six and then eight equal parts (see p. 126, Fig. 14.6). At $H/8$ from the top and $H/4$ from the bottom, draw horizontal lines.
3 At $H/6$, bisect the angles marked C and D to obtain the intersection point E, which should lie on the vertical construction line as shown. Transfer the point E horizontally to the corresponding positions for adjacent teeth.
4 With centres E, draw tangential root arcs and complete the required thread profiles by drawing the thick outlines.

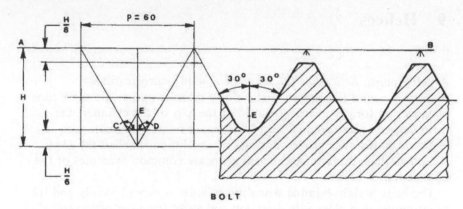

Fig. 8.11 ISO metric external thread

(b) To draw an *internal* (nut) thread M64 × 6 to a scale of 10:1 (Fig. 8.12):

1 From a horizontal line AB, draw say seven vertical lines half a pitch (30 mm) apart and then draw the flanks inclined at 30° to give the thread angle of 60°.
2 Join the apex points to obtain the height H and divide this into eight equal parts (see p. 126, Fig. 14.6). At $H/8$ from the top and $H/4$ from the bottom, draw horizontal lines.
3 Complete the required thread profile by drawing the thick outlines.

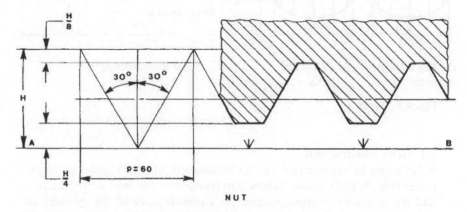

Fig. 8.12 ISO metric internal thread

9 Helices

A helix (plural *helices*) is the locus of a point which revolves uniformly round the curved surface of a cylinder and at the same time advances uniformly in the direction of the axis of the cylinder, the ratio between these angular and linear movements being constant.

The axial distance moved during one revolution is called the *lead*.

Coil springs, threads, and worm gears are common examples of the application of helices.

The helix is right-handed when the cylinder is viewed axially and the point moves in a clockwise direction and away from the observer, Fig. 9.1(a).

The helix is left-handed when the point moves in an anticlockwise direction and away from the observer, Fig. 9.1(b).

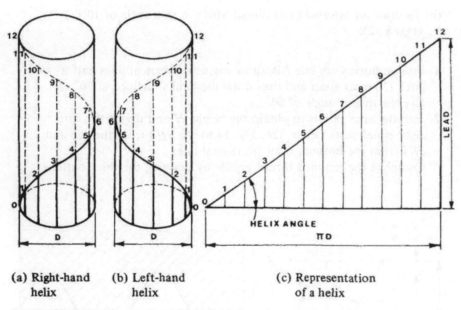

(a) Right-hand helix   (b) Left-hand helix   (c) Representation of a helix

Fig. 9.1 Helix

9.1 Helix construction

A helix can be represented by the hypotenuse of a triangular piece of paper (Fig. 9.1(c)) whose height corresponds to the lead of the helix and whose base is wound around the circumference of the cylinder as shown in Figs. 9.1(a) and (b).

To construct a helix (Fig. 9.2)

1 Draw the front and plan views of the base cylinder.
2 On the plan view, mark say twelve equally spaced points, numbering them anticlockwise for a right-hand helix and clockwise for a left-hand helix as shown.
3 From these points, project vertical lines from the plan to the front view.
4 Divide the lead into the same number of equal parts as the plan (twelve) and number the points as shown.
5 Project horizontally from these points to intersect the corresponding vertical projection lines at points P_1, P_2, ..., P_{12}.
6 With the help of a French curve, join all the intersection points P_1, P_2, ..., P_{12} with a smooth curve, This is the required helix.

To draw a screw thread a double helix is constructed, the larger helix representing the crest of the thread and the smaller one the root of the thread as shown in Fig. 9.2(b).

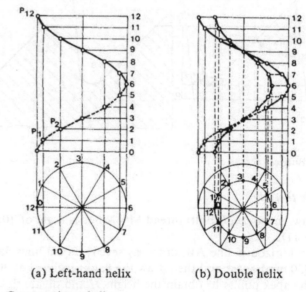

(a) Left-hand helix   (b) Double helix

Fig. 9.2 Constructing a helix

To draw a vee thread (Fig. 9.3)

1 Draw the front view of the construction cylinder, mark off the lead and divide it into, say, twelve equal parts numbered as shown.
2 Draw the circle for the end view of the construction cylinder, corresponding to the crest of the thread.

102

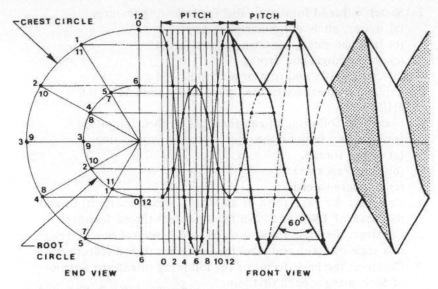

Fig. 9.3 Right-hand vee thread

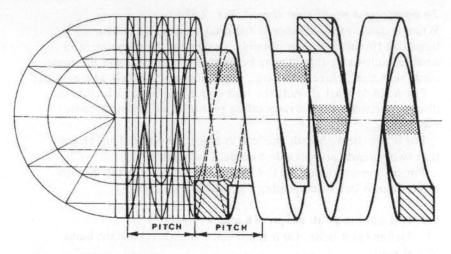

(a) Right-hand square thread   (b) Left-hand square-section spring

Fig. 9.4 Square thread and square-section spring

3 Draw the thread angle of 60° on the front view and project the root of the thread horizontally to the end view.

4 Draw the root circle and divide this and the crest circle into twelve parts as shown. Number the intersection points around the circumference, anticlockwise for a right-hand thread and clockwise for a left-hand thread.

5 From the crest circle, project these points horizontally on to the front view and mark the points where the projector from a particular number on the end view intersects the vertical line with the same number on the front view. These points give the larger helix.

6 Repeat step 5 from the root circle to give the points for the smaller helix, starting from the required point.

7 Complete the front view by drawing smooth curves through the visible parts.

By marking off and dividing several leads along the front view of the construction cylinder and projecting horizontally from the end view through all of them, a greater length of thread may be drawn.

To draw a square thread (Fig. 9.4(a))

The same procedure as given above for the vee thread is followed, except that a square of side equal to the half-pitch and the corresponding helices from four corners are drawn. By choosing the correct helix, a right-hand or a left-hand thread may be obtained.

To draw a square-section spring (Fig. 9.4(b))

Most coiled springs are formed on cylinders, therefore a square-section spring is drawn using the method given above for a square thread but with the cylinder removed, as shown.

To draw a round-section spring (Fig. 9.5)

1 Draw the helix corresponding to the centre line of the round spring wire.

2 With centres on the helix, draw a number of circles of the same diameter as the spring wire.

3 Draw the best smooth curves tangential to these construction circles.

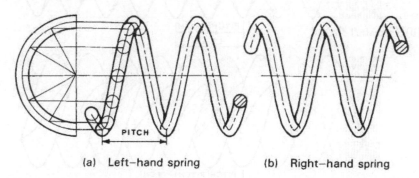

(a) Left–hand spring   (b) Right–hand spring

Fig. 9.5 Round-section spring

103

To construct a multi-start thread (Fig. 9.6)

When a quick axial advance is required, instead of machining a large-size thread and thus possibly weakening the component, a smaller multi-start thread may be used. Such threads can be single-start, two-start, three-start, etc., depending on the helix angles used.

For a single-start thread, the lead equals the pitch, which is the distance between the corresponding points on adjacent threads, Fig. 9.6(a).

For a two-start thread, the lead is twice the pitch. This means that two threads are cut side by side, Fig. 9.6(b).

For a three-start thread, the lead is three times the pitch, with three threads cut side by side, Fig. 9.6(c); and so on.

9.2 Test questions on chapters 8 and 9

1 Define (a) a helix, (b) a multi-start thread, (c) a right-hand thread.

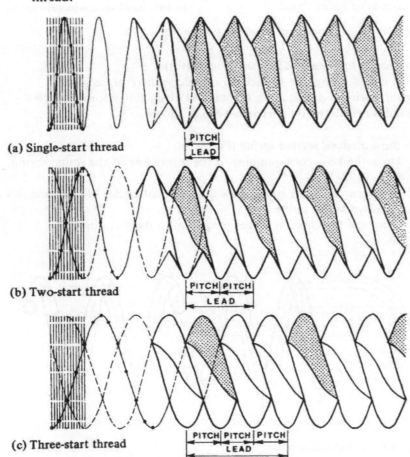

(a) Single-start thread

(b) Two-start thread

(c) Three-start thread

Fig. 9.6 Multi-start thread

2 Sketch a thread form and label for following features:
(a) major, pitch and minor diameters;
(b) lead and pitch, for a single-start thread;
(c) thread angle and depth;
(d) crest, flank, and root of a thread.

3 Explain the meaning of the thread designation M5 × 0.8 – 6H/6g.

4 Sketch the following thread forms, label their important features, and indicate their general use in engineering:
(a) a BSP thread,
(b) a buttress thread,
(c) a square thread.

5 Construct the basic form of an M5 × 0.8 external (bolt) thread to a scale of 100:1 and label the important thread features.

6 Construct the basic form of an M4 × 0.7 internal (nut) thread to a scale of 100:1 and label the important thread features.

7 Construct the basic forms of the following threads to a scale of 50:1, using a pitch of 1 mm:
(a) a BSP thread,
(b) an ISO metric trapezoidal screw thread,
(c) a buttress thread.

8 Explain the difference between a pitch and a lead.

9 Construct three complete turns of a right-hand single-start vee thread to a scale of 10:1 for a pitch of 3.6 mm major diameter of 11 mm, and a thread angle of 60°. (See p. 103, Fig. 9.3).

10 Construct four complete turns of a right-hand two-start square thread with a pitch of 36 mm and a major diameter of 108 mm.

11 Construct four complete coils of a left-hand square-section spring with a pitch of 36 mm and a pitch diameter of 96 mm. (See p. 103, Fig. 9.4).

12 Construct six complete coils of a right-hand round-section spring. Diameter of the spring wire to be 10 mm, pitch (effective) diameter 100 mm, and pitch 48 mm.

13 Construct six complete turns of a right-hand three-start vee thread to a scale of 10:1 for a pitch of 3.6 mm, a major diameter of 12 mm, and a thread angle of 60°.

14 Construct six complete turns of a left-hand three-start square thread with a pitch of 36 mm and a major diameter of 100 mm.

15 Construct five complete turns of a left-hand two-start ISO metric trapezoidal screw thread to a scale of 10:1 for a pitch of 4 mm and a major diameter of 12 mm. Thread depth to be equal to half a pitch, and thread angle to be 30°.

10 Fasteners

10.1 Temporary fastenings

There are different types of fastening to join one part to another. The design and the function of the secured parts must be taken into consideration before finally deciding on the fastening method to be adopted.

There are two classes of fastenings: temporary and permanent.

Temporary fastenings can be used more than once. Any assembled components held together can be dismantled and re-assembled many times without damaging the fastenings.

Nuts and bolts

The bolt has an external thread which extends along only part of the shank. Bolts generally pass completely through the work to be fastened and on the other side are secured by a nut, which has an internal mating thread (see Fig. 10.5(d)). Nuts and bolts are usually hexagonal-headed and are adjusted with a standard spanner of the open-ended, ring, or socket type.

Nuts and bolts often have to be drawn by draughtsmen, so it is very useful to learn a quick method to obtain an approximate shape.

Approximate method of drawing a hexagonal nut (Fig. 10.1)

1. Start with the plan view. Draw a circle of diameter 2*d*, where *d* is the major diameter of the thread (nominal size).
2. Using a 60° set square, construct a hexagon inside the circle and then draw a chamfer circle inside the hexagon.
3. Complete the plan by drawing concentric circles representing the threaded hole.
4. Project the front and end views, making the height of the nut equal to 0.8*d*.
5. From the points marked A in the front view, draw construction lines at 30° to the main centre line to intersect this centre line at point B.
6. With the centre at point B, draw the chamfer curve tangential to the top surface of the nut.
7. From the points marked C, draw construction lines at 30° to the main centre line, intersecting the initial construction line at the points E.
8. With centres at points E, draw two small chamfer curves tangential to the top surface of the nut.

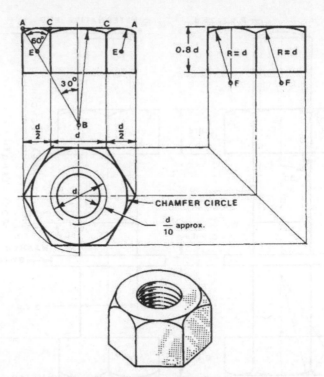

Fig. 10.1 Method of drawing a hexagonal nut

9. In the end view, bisect the distances between the main centre line and the two extreme sides. Starting at the top surface of the nut, mark along the bisectors a distance equal to *d*, giving the intersection points F.
10. With centres at points F, draw chamfer curves tangential to the top surface of the nut.
11. Complete the views, noting the rectangular outline of the end view.

Figure 10.3 shows the completed views of a nut and bolt. Note the nut and bolt head heights and the full-thread and bolt lengths.

Nuts, thin washers, bolts, studs, and screws should not be hatched when sectioned longitudinally, i.e. along their axis.

INT DIA = d + 1 mm
EXT DIA = 2 d
THICKNESS = 0.2 d

Fig. 10.2 Washer proportions in terms of the bolt diameter *d*

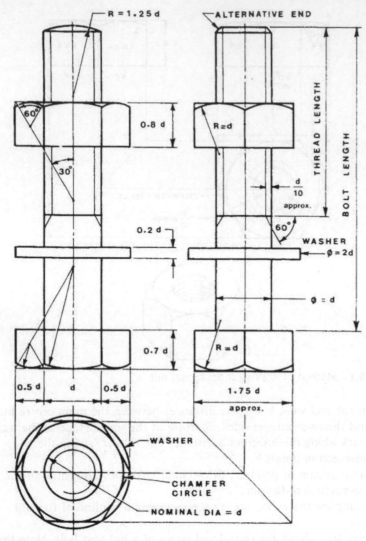

Fig. 10.3 Completed views of a nut and bolt showing thread run-outs

Screws

A screw has an external thread extending almost the whole length of the screw shank to the head.

Screws, like bolts, are used for fastening two or more parts together. One of the parts has a tapped hole and the other part has a clearance hole (see Fig. 10.5(e)). The screw is used by passing it through the clearance hole in one part to screw into the threaded hole in the other, so fastening both parts securely together.

Screws are not secured by a nut.

Apart from hexagonal heads, the following types of screw-head are the most regularly used:

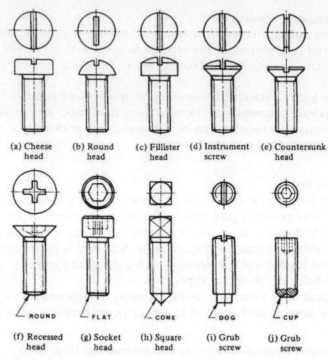

(a) Cheese head (b) Round head (c) Fillister head (d) Instrument screw (e) Countersunk head

(f) Recessed head (g) Socket head (h) Square head (i) Grub screw (j) Grub screw

Fig. 10.4 Various types of screw

Cheese-head (a), *round-head* (b) and *fillister-head* (c) screws (Fig 10.4) are used regularly in engineering, whereas *instrument screws* (d) are used in instrument work. *Countersunk-head screws* (e) are used where a flush surface has to be maintained. *Recess-head screws* (f) need a special cross-shaped screwdriver for tightening. *Socket-head screws* (g) can be placed in counterbored holes to maintain a flush surface. They are tightened by means of a hexagonal wrench and are mainly used in tool-making. *Square-head set screws* (h) and headless *grub screws* (i) and (j) are used to prevent relative rotation or sliding movement between two components.

Flat pointed (g) or *cup-pointed* (j) screws are used where a contact with flat surfaces is required to prevent movement. *Cone-pointed screws* (h) bite into shafts, and *dog-pointed screws* (i) usually fit into slots.

Self-tapping screws (not shown) form their own thread when tightened.

Studs

Studs are threaded on both ends, with an unthreaded shank in the middle, and are used for parts that must be removed frequently, like cylinder heads, covers, lids, etc.

Studs are screwed tightly into tapped holes in the permanent part, while the removable part has clearance holes in the corresponding positions. Nuts are used on the projecting ends of the studs to secure the two parts together, as shown in Fig. 10.5(c).

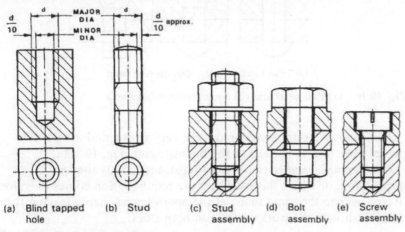

(a) Blind tapped hole (b) Stud (c) Stud assembly (d) Bolt assembly (e) Screw assembly

Fig. 10.5 Studs, bolts and screws

Normally only one washer is used for an assembly of a bolt, stud, or screw. The washer should be placed under the component which is turned in order to tighten the assembly, as shown in Figs 10.5(c) and (d).

Pins

Clevis pin (Fig. 10.6(a))   A clevis or dowel pin is a headless cylindrical pin used for precise-location purposes.

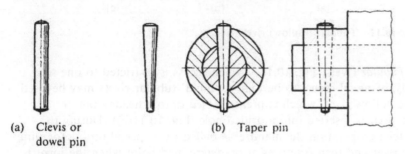

(a) Clevis or dowel pin

(b) Taper pin

Fig. 10.6   Pins

Taper pin (Fig. 10.6(b))   This type of pin is conical with a slight taper. It is usually used to attach cotters, wheels, etc. to shafts. It is forced tightly into a reamed hole having the same taper, which is standardised.

Split cotter pin   The split pin is usually inserted through holes and slots and its ends are opened up as in Fig. 10.7(a).

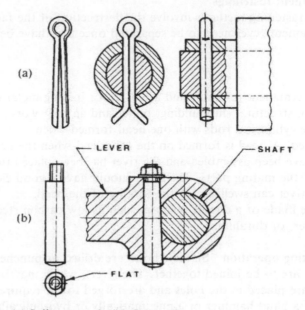

(a)

LEVER — — SHAFT

(b)

— FLAT

Fig. 10.7   (a) Split cotter pin (b) Cotter pin

Cotter pin   The cotter pin is a round rod threaded at one end, or it may be plain with a tapered flat machined along its length. This type is used to secure levers, cranks, etc. to spindles, Fig. 10.7(b). A typical application is attaching a bicycle-pedal crank to the chainwheel spindle.

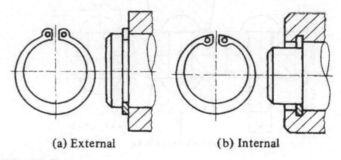

(a) External   (b) Internal

Fig. 10.8   Circlips

Circlips

A component can be located axially against a shoulder of a shaft by a circlip or snap ring as in Fig. 10.8. To provide axial location between a shaft and its bearing, a locating groove is machined and the circlip is sprung into position. Circlips may be internal or external.

10.2 Permanent fastenings

Permanent fastening methods involve the destruction of the fastening if the components ever need to be separated once they have been joined.

Rivets

Rivets are permanent fasteners and are used for joining metal sheets and plates in structural shipbuilding, boiler and aircraft work, etc.

Rivets are cylindrical rods with one head formed when manufactured. A head is formed on the other end when the parts to be joined have been assembled and the rivet has been placed through the holes of the mating parts. The holes should have a good clearance, so that the rivet can swell during the forging of the head.

Rivets are made of a ductile material such as low-carbon steel, brass, copper, or duralumin.

Simple riveting operation Suitable holes are drilled or punched in the parts which are to be joined together. The rivets, which may be cold or heated, are placed in the holes and are forged to the required shape using either a hand hammer or a pneumatically or hydraulically operated gun (Fig. 10.9).

The different types of rivet are defined according to the shape of the rivet head.

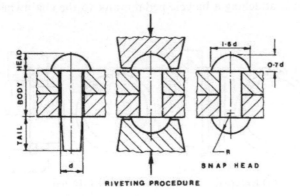

Fig. 10.9 Snap-head rivet

Snap- or round-head rivets are the most commonly used and are easy to shape. These rivets are used mainly for machine riveting and structural work and are riveted over to the same shape at the opposite end, Fig. 10.9.

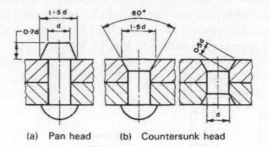

(a) Pan head   (b) Countersunk head

Fig. 10.10 (a) Pan-head and (b) countersunk-head rivets

Pan-head rivets are considered to be very strong and are usually riveted over to a snap head at the opposite end, Fig. 10.10(a).

Countersunk-head rivets, Fig. 10.10(b), are mainly used in shipbuilding or where flush surfaces are required. Sometimes they are riveted over to the same shape at the opposite end, hence leaving the rivets flush with the work surface on both sides.

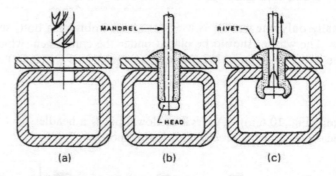

(a)   (b)   (c)

Fig. 10.11 Tubular (hollow) rivet

Tubular rivets, Fig. 10.11 — when access is restricted to one side only, of metal sheets to be joined, 'blind' tubular rivets may be used. The hollow rivet, which is preassembled onto a headed pin or mandrel, is inserted into a drilled hole (Fig. 10.11(b)). During the fastening operation the mandrel is pulled by a special tool deforming the rivet and then fractures at predetermined point when the joint is tight (Fig. 10.11(c)).

Welding

Welding is a process of uniting two pieces of metal by fusing them together to form a permanent joint. This may be done with or without additional (filler) metal and with or without the application of pressure.

Fusion welding In this process where the areas to be joined are heated until they become plastic (i.e. able to flow) and are then welded together with or without the addition of filler metal.

Gas welding is a fusion process in which the heat is provided by burning a gas mixed with oxygen to create a hot flame which is applied to the joint by means of a torch. The most commonly used gases are acetylene, propane, and hydrogen.

Electric-arc welding is a fusion process in which a local area of intense heat is created by passing an electric current through a filler rod, which acts as an electrode, held at a short distance from the joint so that the electric circuit is completed by arcing.

Pressure welding In this process the areas to be joined are heated until they become plastic and are then welded together by applying pressure or sometimes hammering.

In *forge welding*, the pieces to be joined are heated and then hammered together.

In *resistance welding*, the pieces to be joined are butted together under pressure and a heavy current is passed through them, producing sufficient heat for welding under continual pressure.

Spot welding is a process of pressure resistance welding in which thin parts are overlapped and welds are made at successive single spots by electrodes contacting both sides of the metal sheets under pressure.

Welding symbols

On engineering drawings, welding requirements are made clear and unambiguous by using welding symbols specified in British Standard BS 499. These symbols give instructions as to the type of welds, their position, and sometimes their size.

The type of weld to be made is indicated by the type of weld symbol. Some of the welds commonly used and their symbols are shown in Fig. 10.12.

Type of weld	Symbol	Illustration	Symbolic representation
Square butt weld	II		
Single V-butt weld	V		
Double-V butt weld	X		
Single-V but weld with broad root race	Y		
Single-U butt weld	Y		
Double-U butt weld	X		
Single-J butt weld	⊬		
Single-bevel weld with broad root race	⊬		
Spot weld	O		
Fillet weld	△		
Plug weld (Circular or elongated hole, completely filled)	⊓		
Seam weld	⊖		

109 **Fig. 10.12** Welds and symbols

The method used to indicate a weld on a drawing is shown in Fig. 10.13.

(a) if the weld symbol is below the horizontal reference line, the weld is to be made on the arrow side of the joint;
(b) if the weld symbol is above the reference line, the weld is to be made on the side of the joint away from the arrow, known as the 'other' side;
(c) if the weld symbol is shown above and below the reference line, welds are to be made on both sides of the joint.

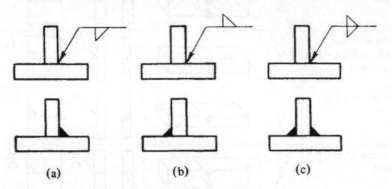

(a) (b) (c)

Fig. 10.13 Weld indication on drawings

A weld all round is indicated by a circle drawn at the elbow of the arrow, Fig. 10.14(a), and a weld to be made on site is indicated by a filled-in triangle as in Fig. 10.14(b).

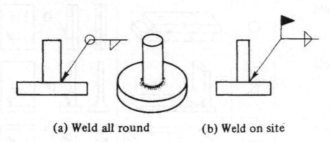

(a) Weld all round   (b) Weld on site

Fig. 10.14 Further welds

If a weld requires special preparation, the arrowhead should always point to the edge of the plate.

10.3 Locking devices

In machinery which is subject to constant working vibration, nuts and bolts tend to work loose – often with disastrous results.

Many methods of locking nuts and bolts have been used, and these fall into two basic groups:

(a) friction locking, which relies on increased friction between the nut and the bolt;
(b) positive locking, which relies on the use of a pin through the nut, various locking plates, washers, or wire locks.

Friction locking devices

Friction locking is usually required on machines which are subject to light vibration. (Friction is the resistance to motion of one surface relative to another surface in contact).

Lock nut Two nuts screwed firmly one on top of the other strain against each other and wedge the nut threads on to the opposite flanks of the bolt threads. This has a locking effect, Fig. 10.15(a).

The thinner lock nut is sometimes fitted beneath the main nut, which carries all the tensile load, and sometimes on top, thus eliminating the need for a thin spanner to adjust it.

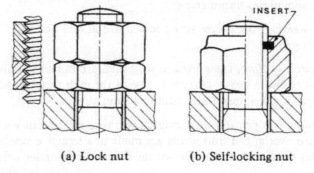

(a) Lock nut   (b) Self-locking nut

Fig. 10.15 Locking nuts

Self-locking nuts These contain a fibre or plastics ring insert, Fig. 10.15(b), which is compressed against the bolt thread when the nut is tightened, thus producing frictional gripping action.

Plain washer A plain washer is not a locking device but provides a flat base for a nut and stops the nut digging in. It is useful on rough surfaces and slots, see p. 105.

Spring washer   When a nut is tightened on to a spring washer, Fig. 10.16(a), the washer's spring action pushes the nut threads against the bolt threads, so increasing the frictional forces.

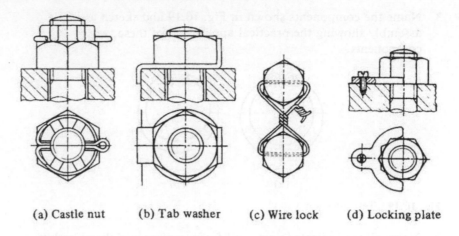

(a) Castle nut   (b) Tab washer   (c) Wire lock   (d) Locking plate

Fig. 10.17   Various locking devices

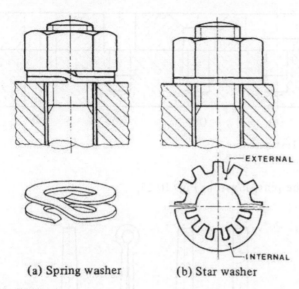

(a) Spring washer   (b) Star washer

Fig. 10.16   Washers

Star washer   These serrated washers have sharp corners, Fig. 10.16(b), which dig into the nut, producing a wedging action and thus forming a semipositive lock.

Positive locking devices
Positive locking devices do not rely on friction and are usually required on machines which are subject to heavy vibrations.

Castle nut   A castle nut is a hexagonal nut with a cylindrical upper part in which slots are cut in line with the centre of each hexagonal face. A hole is drilled through the bolt to correspond with a pair of slots, then a split pin is inserted and the ends are bent over, Fig. 10.17(a). Positive locking is achieved without reducing the number of full threads engaged with the bolt.

Tab washer   The tab washer is inserted between the component and the nut. After the nut has been tightened, one tab is bent closely against a flat of the nut, to prevent rotation, and the other tab is bent down into a drilled hole or over the edge of the component, Fig. 10.17(b).

Wire locking   In this method, soft wire is passed through holes drilled in adjacent bolt heads after the bolts have been tightened. The ends of the wire are twisted together with pliers. The wire must be so arranged that if a nut tends to slacken it should cause the wire to tighten, Fig. 10.17(c).

Locking plate   A locking plate is screwed down so that its jaws engage with the nut and prevent it turning, Fig. 10.17(d).

10.4 Test questions
1   Explain very briefly why washers are used in engineering.
2   Give the correct name for each of the washers shown in Fig. 10.18. Which of them is used as a positive locking washer?

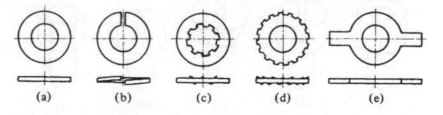

(a)   (b)   (c)   (d)   (e)

Fig. 10.18   Test question 2

3 Name the components shown in Fig. 10.19 and sketch a simple assembly showing the practical application of these two components.

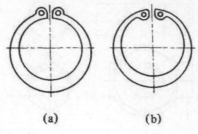

(a)   (b)

Fig. 10.19   Test question 3

4 Name all the indicated parts and features shown in the assembly in Fig. 10.20.

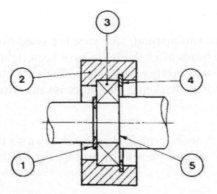

Fig. 10.20   Test question 4

5 (a) What is the main difference between a screw and a bolt?
  (b) What is the main difference between a bolt and a stud?
6 Identify the screw heads shown in Fig. 10.21 by their correct names and suggest the tools used in fastening each of them.

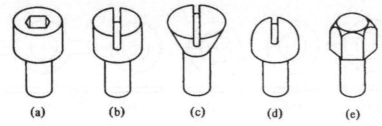

(a)   (b)   (c)   (d)   (e)

Fig. 10.21   Test question 6

7 Draw or sketch the fastenings indicated at (a), and (b), (c) and (d) in Fig. 10.22 for joining two parts temporarily together. Name each fastener.

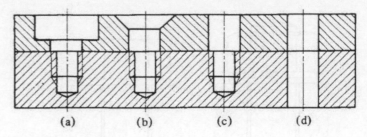

(a)   (b)   (c)   (d)

Fig. 10.22   Test question 7

8 Name the pins shown in Fig. 10.23.

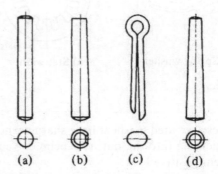

(a)   (b)   (c)   (d)

Fig. 10.23   Test question 8

9 The four plates shown in Fig. 10.24 are not to be machined before welding. Sketch the sections of the most suitable types of weld to join these four plates together, and name each type chosen.

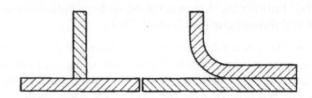

Fig. 10.24   Test question 9

10 Name the types of weld shown in Fig. 10.25 and sketch the appropriate BS welding symbols inside the boxes provided.

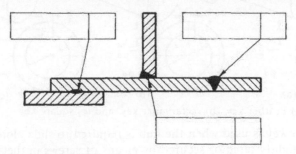

Fig. 10.25 Test question 10

11 By positioning the appropriate symbols on the reference lines, show how the welds in Fig. 10.26 are indicated on a drawing.

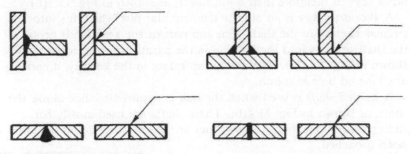

Fig. 10.26 Test question 11

12 State the correct names of the locking devices shown in Fig. 10.27. Which of them is a positive locking device?

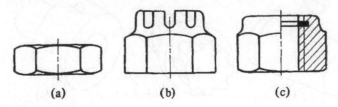

(a)   (b)   (c)

Fig. 10.27 Test question 12

13 State the correct names of the rivets shown in Fig. 10.28. Which rivets are used for (a) structural work? (b) shipbuilding?

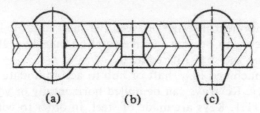

(a)   (b)   (c)

Fig. 10.28 Test question 13

14 State the meaning of the welding symbols shown in Fig. 10.29(a), (b), and (c).

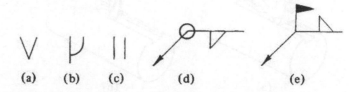

(a)   (b)   (c)   (d)   (e)

Fig. 10.29 Test question 14

15 Describe very briefly the meaning of the welding instructions indicated in Fig. 10.29(d) and (e).
16 Explain the meaning of the terms temporary fastenings' and 'permanent fastenings' and name three examples of each.
17 State the meaning of the terms 'friction locking devices' and 'positive locking devices' and sketch three examples of each type.
18 Redraw, twice full size, the nut and bolt shown in Fig. 10.30 and add a plan.

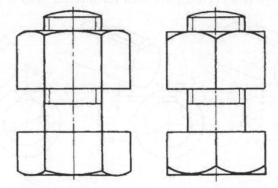

Fig. 10.30 Test question 18

11 Keys

A key is a component inserted between the shaft and the hub of a pulley, wheel, etc., to prevent relative rotation but allow sliding movement along the shaft, if required.

The recess machined in a shaft or hub to accommodate the key is called a *keyway*. Keyways can be milled horizontally or vertically, as shown in Fig. 11.1. Keys are made of steel, in order to withstand the considerable shear and compressive stresses caused by the torque they transmit.

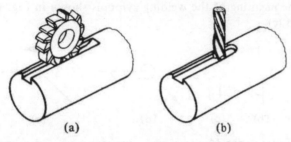

Fig. 11.1   Keyways milled (a) horizontally, (b) vertically

There are two basic types of key:

(a) *Saddle keys*, which are sunk into the hub only. These keys are suitable only for light duty, since they rely on a friction drive alone.
(b) *Sunk keys*, which are sunk into the shaft and into the hub for half their thickness in each. These keys are suitable for heavy duty, since they rely on positive drive.

    Hollow saddle keys (Fig. 11.2(a)) are used for very light duty.
    Flat saddle keys (Fig. 11.2(b)) are used for light duty.
    Round keys (Fig. 11.2(c)) are used for medium duty.

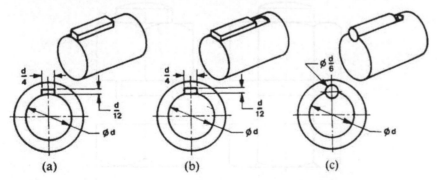

Fig. 11.2   (a) Hollow saddle key, (b) flat saddle key, (c) round key

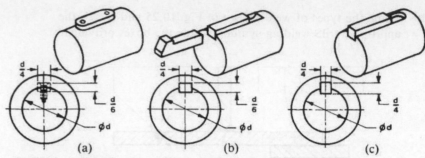

Fig. 11.3   (a) Feather key, (b) rectangular key, and (c) square key

A *feather key* is used when the hub is required to slide along the shaft. It is tightly fitted or secured by means of screws in the shaft keyway, and is made to slide in the hub keyway, as shown in Fig. 11.3(a).

Rectangular and square keys can be parallel or tapered with a basic taper of 1 in 100 to prevent sliding. These keys are used for heavy-duty applications. Students are advised to use square keys for assembly-drawing solutions. *Gib heads* are sometimes provided on taper keys to facilitate their withdrawal, as shown in Fig. 11.3(b).

A *Woodruff key* is an almost semi-circular disc which fits into a circular keyway in the shaft. The top part of the key stands proud of the shaft and fits into the keyway in the parallel or tapered hub, as shown in Fig. 11.4(a). As the key can rotate in the keyway, it can fit any tapered hole in a hub.

A *splined shaft* is used when the hub is required to slide along the shaft, as shown in Fig. 11.4(b). These shafts are used mostly for sliding-gear applications. The splines are usually milled and the splined holes broached.

Square-head set screws and *grub screws* are also used for low-torque applications as shown in Fig. 11.4(c) and also 10.4(h), (i), and (j), p. 106.

If the torque to be transmitted is too great for one grubscrew or key, two may be used set at 90° to 120° around the shaft, but never at 180°.

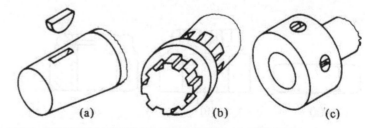

Fig. 11.4   (a) Woodruff key, (b) splined shaft, and (c) grub screws

12 Engineering diagrams

Engineering drawings represent items that can be seen or visualised. Usually they are drawn to some scale and dimensioned with true values according to British Standard 308.

Diagrams

Engineering diagrams usually indicate only the relative positions of interconnected components or systems represented by their relevant symbols, according to British Standard 5070.

Block diagrams indicate, in a simplified form, a functional system where a number of blocks represent the elements of that system. They are interconnected by the main paths of signal or flow of fluid. (Fig. 12.1).

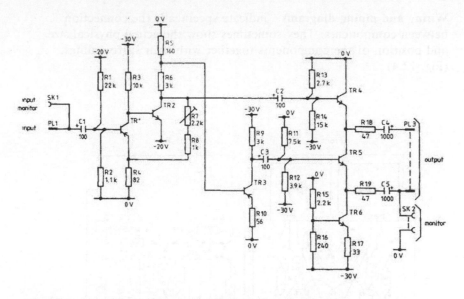

Fig. 12.2 Example of part of circuit diagram: video amplifier

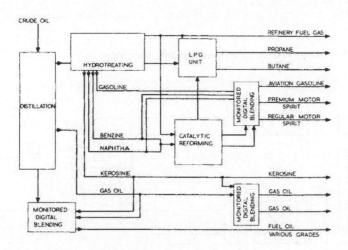

Fig. 12.1 Example of block diagram: refinery link-up

Circuit diagrams indicate the standard symbols representing the functional components and their connections disregarding their actual physical size or position. (Fig. 12.2 and 12.3).

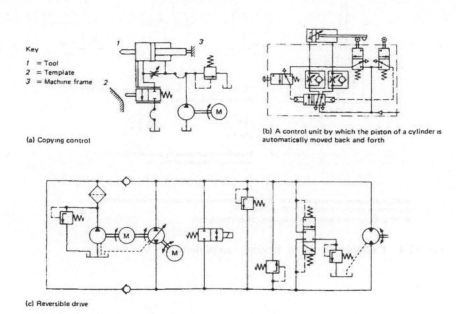

Fig. 12.3 Example of composite diagrams of pneumatic and hydraulic symbols

115

Wiring and piping diagrams indicate specifically the connection between components. They sometimes show the actual physical size and position of the components together with their surroundings. (Fig. 12.4)

Installation diagrams indicate the connections relative to a building, structure, etc. They usually include the relevant full installation information. (Fig. 12.5).

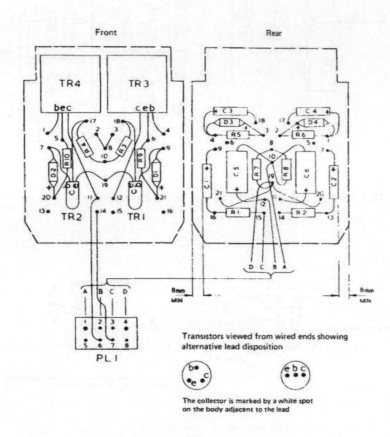

NOTE 1. All component references and terminal numbers are to be clearly marked in indelible characters on the chassis adjacent to the components.

NOTE 2. All transistor leads to be maintained 25 mm min. length.

Fig. 12.4 Example of wiring diagram: amplifier unit

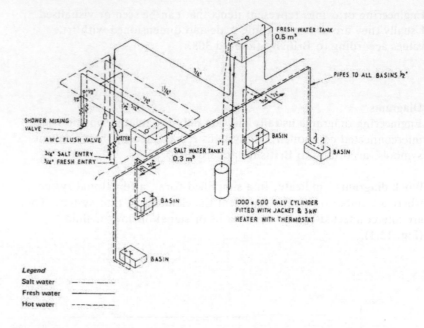

Fig. 12.5 Example of installation diagram: plumbing

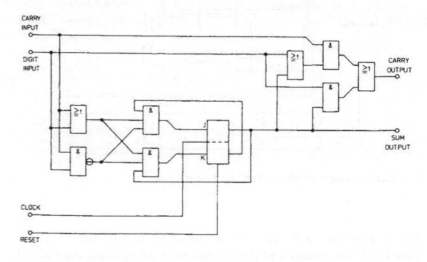

Fig. 12.6 Example of logic diagram: one-stage accumulator

116

Table 12.1 Selected electrical and electronic graphical symbols (from BS 3939)

Description	Symbol	Description	Symbol	Description	Symbol	Description	Symbol
Direct current		Fuse		Galvanometer		Photodiode	
Alternating current		Resistor, general symbol		Oscilloscope		Light-emitting diode, LED	
Positive polarity		Variable resistor		Motor			
Negative polarity		Resistor with sliding contact		Generator		Tunnel diode	
Variability		Potentiometer with moving contact		Microphone		PNP transistor	
Pre-set adjustment		Heating element		Earphone (receiver)		NPN transistor with collector connected to envelope	
Primary or secondary cell		Capacitor, general symbol		Transducer head, general symbol		Amplifier, simplified form	
Battery of primary or secondary cells		Polarized capacitor		Loudspeaker		Laser	
Alternative symbol		Voltage-dependent polarized		Electric clock			
Earth or ground		Capacitor with pre-set adjustment		Make contact, normally open		Binary logic elements	AND
Signal lamp, general symbol		Inductor, winding, coil, choke		Break contact, normally closed			OR
Electric bell		Inductor with magnetic core		Make contact with spring return			Logic identity
Electric buzzer		Transformer with magnetic core					Exclusive OR
Crossing of conductors with no electrical connection		Ammeter		Break contact with spring return			NOT (negator)
Junction of conductors		Voltmeter		Relay			NAND
Double junction of conductors		Wattmeter					
Plug (male)		Watt-hour meter		Semiconductor diode, general symbol			NOR
Socket (female)							

117

Table 12.2 Selected symbols for fluid power systems (from BS 2917)

NOTE 1. The symbols for hydraulic and pneumatic equipment and accessories are *functional* and consist of one or more *basic symbols* and in general of one or more *functional symbols*. The symbols are neither to scale nor in general orientated in any particular direction.

NOTE 2. In circuit diagrams, hydraulic and pneumatic units are normally shown in the unoperated position.

NOTE 3. The symbols show connections, flow paths and the functions of the components, but do not include constructional details. The physical location of control elements on actual components is not illustrated.

Description	Symbol	Description	Symbol	Description	Symbol
General symbols		Air-oil actuator (transforms pneumatic pressure into a substantially equal hydraulic pressure or vice versa)		spring loaded (opens if the inlet pressure is greater than the outlet pressure and the spring pressure)	
Basic symbols					
Restriction:		**Directional control valves**		pilot controlled (opens if the inlet pressure is higher than the outlet pressure but by pilot control it is possible to prevent:	
affected by viscosity		*Flow paths:*			
unaffected by viscosity		one flow path			
Functional symbols		two closed ports			
hydraulic flow		two flow paths		closing of the valve	
pneumatic flow or exhaust to atmosphere		two flow paths and one closed port		opening of the valve)	
Energy conversion		two flow paths with cross connection		with restriction (allows free flow in one direction but restricted flow in the other)	
Pumps and compressors		one flow path in a by-pass position, two closed ports			
Fixed capacity hydraulic pump:					
with one direction of flow		*Directional control valve 2/2:*		Shuttle valve (the inlet port connected to the higher pressure is automatically connected to the outlet port while the other inlet port is closed)	
with two directions of flow		with manual control			
Motors		controlled by pressure against a return spring			
Fixed capacity hydraulic motor:		*Direction control valve 5/2:*		**Pressure control valves**	
with one direction of flow		controlled by pressure in both directions		*Pressure control valve:*	
Oscillating motor:		NOTE. In the above designations the first figure indicates the number of ports (excluding pilot ports) and the second figure the number of distinct positions.		one throttling orifice normally closed	or
hydraulic				one throttling orifice normally open	or
Cylinders		**Non-return valves, shuttle valve, rapid exhaust valve**		two throttling orifices, normally closed	
Single acting: Detailed / Simplified					
returned by an unspecified force		*Non-return valve:* free (opens if the inlet pressure is higher than the outlet pressure)		*Sequence valve* (when the inlet pressure overcomes the spring, the valve opens, permitting flow from the outlet port)	
Double acting: with single piston rod					
Cylinder with cushion: single fixed					

Description	Symbol	Description	Symbol	Description	Symbol
Flow control valves		*Rotary connection*		Over-centre device (prevents stopping in a dead centre position)	
Throttle valve: simplified symbol		one way			
		three way		*Pivoting devices:* simple	
Example: braking valve		**Reservoirs**		with traversing lever	
Flow control valve (variations in inlet pressure do not affect the rate of flow): with fixed output	Simplified	Reservoir open to atmosphere:		with fixed fulcrum	
		with inlet pipe above fluid level		**Control methods**	
		with inlet pipe below fluid level		*Muscular control:* general symbol	
Flow dividing valve (divided into a fixed ratio substantially independent of pressure variations)		with a header line		by push-button	
		Pressurized reservoir		by lever	
		Accumulators		by pedal	
Energy transmission and conditioning		The fluid is maintained under pressure by a spring, weight or compressed gas		*Mechanical control:* by plunger or tracer	
Sources of energy				by spring	
Pressure source		**Filters, water traps, lubricators and miscellaneous apparatus**		by roller	
Electric motor		*Filter or strainer*		by roller, operating in one direction only	
Heat engine		**Heat exchangers**		*Electrical control:* by solenoid (one winding)	
Flow lines and connections		*Temperature controller* (arrows indicate that heat may be either introduced or dissipated)		by electric motor	
Flow line: working line, return line and feed line				*Control by application or release of pressure*	
pilot control line		*Cooler* (arrows indicate the extraction of heat)		*Direct acting control:* by application of pressure	
drain or bleed line				by release of pressure	
flexible pipe		with representation of the flow lines of the coolant			
Pipeline junction		*Heater* (arrows indicate the introduction of heat)		*Combined control:* by solenoid and pilot directional valve (pilot directional valve is actuated by the solenoid)	
Crossed pipelines (not connected)		**Control mechanisms**			
Air bleed		**Mechanical components**			
Power take-off plugged		*Rotating shaft:* in one direction			
with take-off line		in either direction		**Measuring instruments**	
connected, with mechanically opened non-return valves		*Detent* (device for maintaining a given position)		*Pressure measurement:* pressure gauge	
uncoupled, with open end		*Locking device* (* symbol for unlocking control)		**Other apparatus**	
uncoupled, closed by free non-return valve				*Pressure electric switch*	

118

13 Computer Aided Draughting (CAD)

The use of computers to produce engineering drawings began in the early 1980s. These systems were both crude and expensive. The technological advances in both hardware and software since the late 1980s have provided users in the engineering industry with increasingly efficient, more reliable and cheaper systems.

13.1 CAD hardware

The hardware requirements for a CAD system can generally be classified as follows:

(a) mainframe systems,
(b) mini systems,
(c) workstations,
(d) personal computers.

Mainframe computers are relatively large and expensive with high memory capability for storing information and necessary instructions. They are capable of performing many tasks simultaneously at great speed and can serve a large number of terminals, which are devices necessary to communicate with the computers. Mainframes are generally used for integrated Computer Aided Design and Draughting (CADD) and also for Computer Aided Manufacture (CAD–CAM).

Mini systems are smaller and cheaper versions of mainframe systems. The recent trend towards miniaturisation has resulted in much increased memory capabilities, inasmuch as they are progressively replacing mainframe systems. *Minicomputers* can solve a variety of complex problems and are capable of producing complicated three-dimensional images. They can also support several terminals.

Workstations are self-contained units comprising a *microcomputer*, usually small and portable, linked to printers, plotters and modems for transmitting data to any destination. Workstations have the ability to work as 'stand-alone units', they have large memory capabilities and they allow limited multi-tasking operations (performing several tasks simultaneously) as well as multi-user capabilities.

Personal computers (PCs) are microcomputers that are becoming increasingly popular for CAD. They are 'stand-alone' systems and can operate with most commercial software. Low cost and increasing speed of operation and reliability have made PCs very popular with CAD users. They generally lack the multi-tasking and multi-user capabilities, although they can be networked to a fileserver.

13.2 What are CAD, CADD and CAM?

The term CAD can represent two entirely different concepts, namely:

(a) Computer aided draughting (drawing),
(b) Computer aided design.

Computer aided draughting is the production of engineering drawings that were traditionally obtained using a drawing board and T-square.

Computer aided design encompasses the concept of using computers to aid the design process. The features that make computers an invaluable aid to the designer include calculation capabilities, analysis, evaluation and modelling. For further information see pages 159–160.

The combination of design and draughting using computers is now being called *Computer Aided Design and Draughting* (CADD).

Computer Aided Manufacture (CAM) is a system that uses computers to assist manufacture mainly involving Computer Numerical Control (CNC) machine tools.

The interaction between the CAD and CAM is provided in the geometric database, which contains all the stored information generated by the drawing procedure to enable the manufacturing process of a designed component.

13.3 CAD system (Fig. 13.1)

The term 'system' refers to a number of different computer elements connected together to produce a useful function. This function may be for stock control, accounting, inventory planning, Computer Aided Draughting (CAD), Computer Aided Manufacture (CAM), etc.

All computer systems consist of four elements, namely:

(a) input devices,
(b) processing element,
(c) storage devices,
(d) output devices.

The actual function of the computer system will determine the nature of the element.

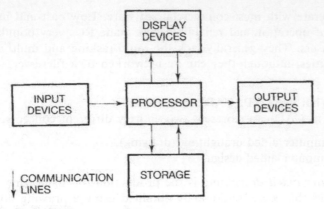

Fig. 13.1 Basic CAD system elements

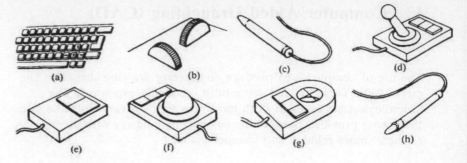

Fig. 13.2 Input devices

Hardware and software

The tasks that are to be performed by a computer system are obtained by combinations of hardware and software.

Hardware is the name given to the physical items in a computer system, e.g. the keyboard, monitor, printer, plotter, disc units, etc.

Software is made up of computer programs, procedures and rules used in the operation of a computer system. A particular computer system may be used for several applications, and each application will have its own software.

For computers to work at all they must have an *operating system*, i.e. a list of instructions in their memory. There are several of these systems, e.g. DOS, UNIX, PRIMOS, AEGIS, VMS, etc.; DOS (Disc Operating System) is generally used for personal computers (PCs).

Input devices

Input devices allow the CAD user to communicate with the computer. They are used for making selections from a *menu*, which is a layout of a variety of commands and functions required to operate the system. Sending these commands into the computer produces complete engineering drawings. The menus may be displayed in a variety of forms such as on the flat surface of a *digitising tablet*, which is connected to the computer, or on the *computer screen* itself, often in the form of *icons*, which are usually the symbolic representations of a facility available.

The choice of optional commands on the screen menu is made by indicating the required position with cursor cross-hairs or by typing the required code.

The movement of the cursor on the screen may be controlled by the devices listed below (see Fig. 13.2).

(a) a *keyboard*, where the allocated keys control the required movements;

(b) *thumb wheels*, where one wheel controls horizontal movement and another wheel controls vertical movement;

(c) a *light pen*, where the required position is selected by pointing the pen directly at the screen;

(d) a *joystick*, where a vertical lever mounted in a box controls movements in any direction;

(e) a *mouse*, a small box which when pushed across a surface controls movements in the same direction;

(f) a *tracker ball*, whose rolling in the mounting controls movements in any direction;

(g) a *puck* or (h) *stylus*, in conjunction with the digitising tablet, can enter the complete drawing from a sketch or half-completed drawing by attaching it to the surface of the digitiser, indicating important points on lines, curves, etc., and entering the relevant commands.

Display

The display devices are used for visually showing information. There are two main types of computer display screen: raster and refresh.

The raster scan tube, which is similar to a television screen, uses a grid of dots known as pixels to display the image. The resolution or clarity of the image depends upon the number of pixels per screen area. The picture is refreshed or scanned very rapidly in regular horizontal sweeps to cover the entire screen at a rate of around 50 scans per second with the intensity varied as necessary for all points that form the picture.

This system achieves a fast update and a good quality of colour and resolution.

The refresh tube system is maintained by regular redrawing, 50 times per second, of the picture lines, arcs and curves. First, the points are located, then the required connecting lines or curves are drawn directly with an electron beam.

The updating is very rapid but colour capability is limited and there is a tendency to flicker as the picture becomes more complex.

Output
The output devices receive data from the computer and provide an output hard copy. There are two main types of output device, namely printers and plotters.

Printers are used for printing out paper copies of programs, data, calculations, etc. They may be impact, where images are formed by a striking action, or non-impact:

(a) *impact printers* for drawing reproduction are of the *dot matrix* type and form shapes by the appropriate selection of small dots from the printhead;
(b) *non-impact printers* include electrostatic, ink-jet and laser printers:

   (i) *electrostatic printers* create shapes by burning away a thin metallic coating on the special printing paper;
   (ii) *ink-jet printers* use a printhead which directs a jet of ink at the paper to create the required shapes;
   (iii) *laser printers* use a fine beam of laser light to create the required shapes.

Plotters are used for drawing graphs, engineering drawings, etc., using one or more pens for writing and drawing on paper. There are two basic types of plotter, i.e. flatbed and drum:

(a) *flatbed plotters* have a flat area on which paper of any type and thickness is placed and pens of various thicknesses and ink colours are free to move in any direction with the plothead providing all the motion;
(b) *drum plotters* have a rotating drum over which the paper can move in two directions; pens are limited to moving across the drum only and, with a combination of pen movement and drum rotation, this provides the required motion.

Storage
There are two memories for storing computer information: symbols, programs, graphics, etc.

(a) *Primary storage* is the internal main memory, which is connected directly to the central processor unit.
(b) *Secondary storage* is the auxiliary memory stored externally on magnetic 'floppy' discs and loaded into the main memory when required.

13.4 CAD software
Buying software for CAD is difficult. There are many commercial packages on the market, and the buyer must try to ensure that the correct package is obtained. If 2-D draughting is the requirement then there is no need to buy a more expensive package with, say, finite element analysis. Some packages are designed for mainframes/workstations and will not run on PCs.

For larger systems, companies such as Computervision, IBM, Intergraph and Microstation provide packages which allow extensive design calculations to be achieved in both wire-frame and solid model format as well as animation.

For PCs there are many packages available, the most popular being AutoCAD. This will allow 2-D draughting, 3-D draughting and solid modelling, which will satisfy most companies' draughting requirements.

It must be remembered that all draughtsmen should have a sound training in producing drawings with traditional drawing instruments before using computers, and they also should be familiar with all relevant British Standards.

There are four basic elements of a drawing: *points*, *straight lines*, *arcs* and *curves*.

The points on the screen can be located by selecting the position with the cursor, snapping to the grid points or numerically using the co-ordinate systems:

(a) *Cartesian co-ordinates* use the horizontal distance x and the vertical distance y from the fixed origin to locate the required point, as shown in Fig. 13.3(a),
(b) *polar co-ordinates* use the length of the radius R and an anticlockwise angle measured from 'east' to locate the required point, as shown in Fig. 13.3(b).

There are many useful functions available in a CAD system which are not possible in manual draughting (see Fig. 13.4);

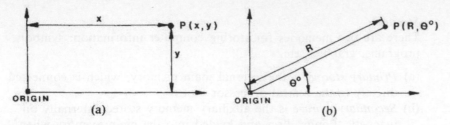

Fig. 13.3 (a) Cartesian co-ordinates, (b) polar co-ordinates

scaling is the capability to enlarge or diminish the size of a displayed feature without changing its shape;

zooming is the capability to enlarge or reduce a selected area of drawing;

rotation is the capability of rotating the features about a selected centre and redrawing them at the new angular position;

mirroring is the capability to create the reverse image of a feature about the chosen line of symmetry;

duplicating is the capability to redraw a feature or component many times and display it in an orderly manner linearly or rotationally. The CAD systems offer the following additional automatic facilities:

fillets: arcs blending together into intersecting lines or curves;

chamfers: surfaces created by bevelling an edge or corner;

tangents: lines constructed to touch circles and arcs precisely;

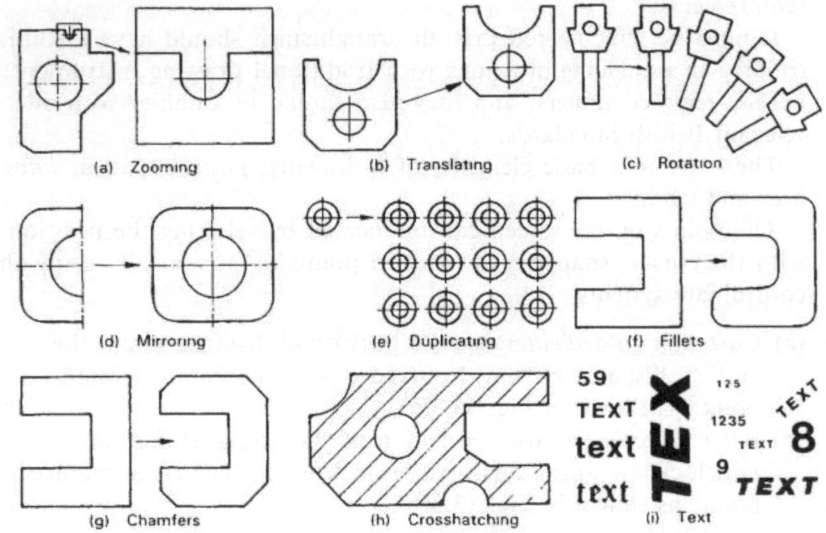

(a) Zooming (b) Translating (c) Rotation

(d) Mirroring (e) Duplicating (f) Fillets

(g) Chamfers (h) Crosshatching (i) Text

Fig. 13.4 Some CAD facilities

crosshatching: shading of an area defined by its boundaries;

dimensioning: the required features are measured by a computer and the correct dimensional value is displayed;

text can be produced in any size and style and can be placed at any angle.

13.5 Benefits of using CAD

(a) *Constant quality drawing*: the quality of lines, dimensions, symbols, notes, etc., is independent of the individual skill of the draughtsman.

(b) *Creation of database*, which is the collection of useful information that may be retrieved by draughtsmen and accessed by other processors.

(c) *Creation of library* of commonly used electrical, hydraulic, welding, etc., symbols for standard components such as nuts, bolts, screws, bearings, etc., projection symbols, parts of drawings, etc. These can be stored in the memory and recalled when needed, positioned anywhere on the screen and redrawn to any scale and angle of inclination.

(d) *Use of layers*: the drawings may be drawn on any one of a number of available layers, which may be considered as a stack of transparent sheets; any separate sheet can be selected for drawing construction lines, grids, dimensions, notes, hatching, etc., or made up together to a complete drawing when required.

(e) *Saving on repetition*: repetitive work on similar features or drawings and the resulting tiredness and boredom is replaced by automatic redrawing; hence attention and interest are maintained with consequent marked increase in speed and productivity.

(f) *Greater accuracy*: because of computer mathematical accuracy, a high level of dimensional control is obtained with subsequent reduction in the number of mistakes; this results in accurate material and cost estimates.

(g) *Multicolour drawings*: visualisation of drawings relates directly to the projection used; pictorial projections are easier to understand than orthographic projections and the different colours available on computer enhance the understanding even further.

(h) *Editing functions*: the powerful editing functions of correcting mistakes, deleting and inserting new features, copying, moving, translating and rotating features, scaling, etc., are only possible with the use of computers.

(i) *modems*: modern communication links, using standards such as IGES, allow drawings to be transferred to users anywhere in the world.

122

(j) *CAD–CAM link*: the drawing can be translated to produce a part program in order to manufacture the component.

(k) *Increased production rate*: a trained draughtsman using a CAD system can produce drawings many times faster than by traditional methods.

While there are many benefits to be gained from installing a CAD system, it should also be noted that problems can arise if the implementation of the system is not carried out effectively. Full consideration should be given by the purchaser to the choice of a system, consultation with employees, training requirements and lead time before the system becomes operational.

13.6 Three-dimensional modelling

A model is a mathematical representation of a geometric form which is stored in the computer memory of a CAD system.

Two-dimensional system

In this system the graphics screen is used as a substitute for drawing paper. Each view of the component must be individually drawn only in one plane without any depth, therefore the computer is unable to automatically generate views additional to those already created.

Three-dimensional system

Most of the larger CAD systems have an ability to model in three dimensions. The spatial image of the object is drawn in a pictorial projection using x–y–z co-ordinate geometry (Fig. 13.5) and is stored in the memory. It can be recalled and redrawn in 3-D pictorial projection or in orthographic projection representing the image of the object in a number of 2-D views, i.e. front, end, plan and auxiliary views.

Wire-frame modelling

A wire-frame model is a 3-D line drawing of an object showing only the edges without any side surface in between. The image of the object, as the name implies, has the appearance of a 'hollow' frame made from thin wires representing the edges, projected lines and curves, as shown in Fig. 13.6(a).

The main disadvantage of this wire-frame representation is that the hidden detail lines are shown and the resulting drawing is a maze of lines, which can be very confusing and disorientating. The model has length, width and depth, but no volume, and is therefore unsuitable for calculation purposes.

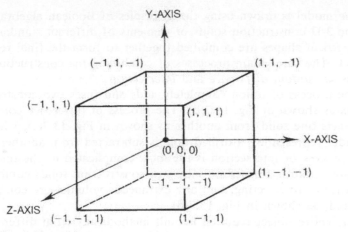

Fig. 13.5 3-D model

Surface modelling

A surface model is more sophisticated than wire-frame representation. Initially the wire-frame is created and the gaps between all individual lines and curves are then 'filled' in by flat or rounded surfaces. The model can be used for surface area calculations, but not for calculations involving volume. The surfaces can be highlighted, modified and colour-shaded. Figure 13.6(b) shows one method of surface representation.

Solid modelling

A solid model allows a true representation of an object (in colour if necessary) displaying full surfaces with the light, highlights and shadows, thus accomplishing very realistic images, as shown in Fig. 13.6(c). The model can be used for calculation purposes to obtain volume, mass, moment of inertia, centre of gravity, etc.

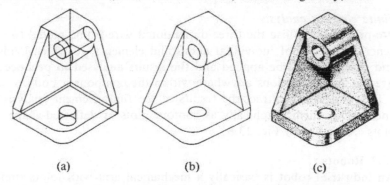

| (a) | (b) | (c) |

Fig. 13.6 (a) Wire-frame model, (b) surface model, (c) solid model

123

The model is drawn using the principles of Boolean algebra.

The 3-D construction solids or elements of different standard geometrical shapes are combined together to form the final required object. The three main processes of combining the construction solids are *union*, *difference* and *intersection*.

The process of union completely adds and fuses two construction solids as shown in Fig. 13.7(b). The process of difference completely subtracts one solid from another as shown in Fig. 13.7(c), where a cylinder representing a drilled hole is subtracted from another solid. The process of intersection is the most complicated of the three. The required solid is formed only by two construction solids combining together or intersecting, and any extraneous volumes are completely ignored, as shown in Fig. 13.7(d).

The entire object required is built methodically from different construction solids in many progressive stages.

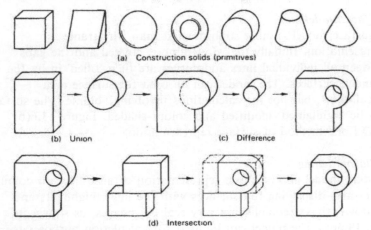

(a) Construction solids (primitives)

(b) Union   (c) Difference

(d) Intersection

Fig. 13.7 3-D modelling

Finite element analysis

Pre-processors utilise the three-dimensional wire-frame model to generate a series of individual nodes and elements, to which loads and restraints can be applied and the results analysed to produce stresses and deflections anywhere within the component body.

Post-processors extract the results of the finite-element analysis and display them graphically as contour plots or deflected-shape plots, as shown in Fig. 13.8.

13.7 Robots

An industrial robot is basically a mechanical arm with joints similar to the human shoulder, elbow and wrist as shown in Fig. 13.9.

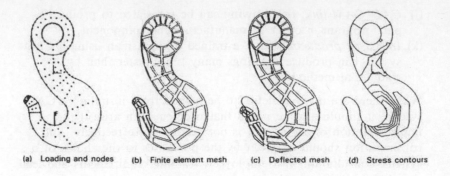

(a) Loading and nodes   (b) Finite element mesh   (c) Deflected mesh   (d) Stress contours

Fig. 13.8 Finite element diagrams (see p. 130, Fig. 14.24(4))

There are six basic movements or degrees of freedom, which provide the robot with the ability to move the *gripper* or *tool* through programmed motions in order to perform a variety of tasks.

The computer enables the robot to perform activities which are based on information stored in a database. This information must be in the correct sequence to ensure the required movements, positions and actions of the robot. To program a task, an operator can guide the arm through the movements needed for a specific operation and these movements are then stored in the computer memory and can be repeated when required.

Depending on the attachment fitted to the gripper, the robot could be machining, welding, spraying paint or assembling, for example. Special attachments may include: vacuum suction cups for lifting flat objects, ladles for liquids, scoops for powders, hooks for lifting purposes and magnetised devices for ferrous objects.

Robots are increasingly used as an integral part of CAD–CAM systems, especially in hostile environments, e.g. extreme temperatures, dust, toxic gases, radioactivity, etc.

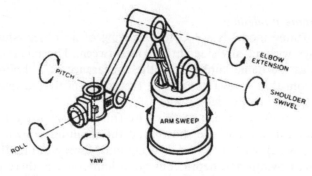

Fig. 13.9 Robot

14 Construction geometry

14.1 Bisectors, fillets and dividers

To draw a line parallel to a given straight line AB at a distance R from it (Fig. 14.1)

1 Set compasses at R and draw two arcs with centres on AB.
2 Draw a tangential line CD to touch both arcs. CD is the required line.

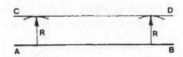

Fig. 14.1 Drawing a parallel line

To bisect a given line AB and to draw a perpendicular to it

1 With centre A and radius greater than half AB draw an arc.
2 With centre B and the same radius draw an arc intersecting the previous arc at C and D.
3 Join CD and this line is perpendicular to and also bisects AB.

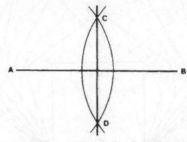

Fig. 14.2 Bisecting a given line and drawing its perpendicular

To bisect a given angle ABC (Fig. 14.3)

1 With centre B draw an arc to intersect BC at E and AB at D.
2 With centres D and E draw arcs of equal radii to intersect at F.
3 Join FB and this line bisects angle ABC.

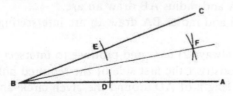

Fig. 14.3 Bisecting a given angle

To draw a fillet, radius R, tangential to given lines AB and CD (Fig. 14.4)

1 Draw line GE parallel to AB and distance R from it.
2 Draw line EF parallel to CD and distance R from it.
3 The intersection of GE and EF gives the centre for the fillet radius, which is drawn between two centre lines EB and EC normal to AB and CD

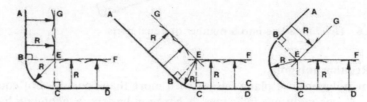

Fig. 14.4 Drawing a fillet with radius R_2 tangential to given lines

To draw a fillet, radius R_2, tangential to an arc and joining a line AB (Fig. 14.5)

1 With centre O and radius $R_1 + R_2$ draw a construction arc.
2 On the side of the required fillet centre, draw a construction line CD parallel to the given straight line AB and distance R_2 from it intersecting the construction arc at D.
3 With centre D and radius R_2 draw the required blending fillet between B and E, where DEO joins two arc centres.

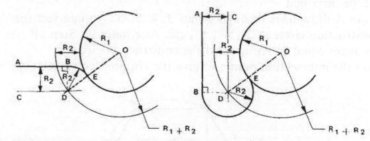

Fig. 14.5 Drawing a fillet with radius R_2 tangential to an arc

To divide a line AB of any length into a number of equal parts (Fig. 14.6)

1 Draw line AC to a convenient length at any angle to AB and divide AC into the required number of equal parts (say 6).
2 Join the last point, 6 on AC to the end of AB and draw lines parallel to 6B through points 1 to 6 by sliding a set square along a straight edge. These lines divide AB into the required equal parts.

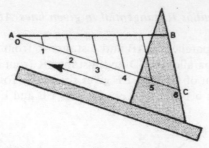

Fig. 14.6 Dividing a line into a number of equal parts

14.2 Regular polygons

Regular polygons are plane figures with more than four sides of equal length. A *pentagon* has five sides, a *hexagon* has six, a *heptagon* has seven, an *octagon* has eight, a *nonagon* has nine and *decagon* has ten sides.

To draw a regular polygon of n sides on a given side AB (Fig. 14.7)
1 Draw the given side AB and produce to C.
2 With centre A and radius AB draw a semicircle between B and C.
3 Divide this semicircle into *n* equal parts numbering each intersection point.
4 Join A to point 2 (always 2).
5 Bisect lines 2A and AB with bisectors intersecting at O.
6 With centre O and radius OA draw a circle in which the polygon will be inscribed.
7 From A draw lines through points 3, 4, 5, etc. to intersect the construction circle at 3', 4', 5', etc. Alternatively: Step off distances equal to AB around the construction circle.
8 Join the intersection points to give the required polygon.

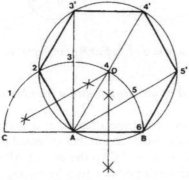

Fig. 14.7 Drawing a regular polygon (hexagon)

To draw a regular polygon of n sides on a given side AB (Fig. 14.8)
1 Draw the given side AB and then draw a bisector EF.
2 At A draw a perpendicular AD equal in length to AB.
3 Draw a diagonal DB to intersect the bisector EF at point 4.
4 With centre A and radius AB draw an arc DB intersecting the bisector at point 6. This is the centre for a circumscribing circle for a hexagon.
5 Bisect the distance between points 4 and 6 on EF to get point 5. This is the centre for a circumscribing circle for a pentagon.
6 On EF step off distances equal to the distance between intersection points 5 and 6. Thus constructed intersection points will be the required centres of circumscribing circles for heptagon at 6, octagon at 8, etc., All drawn on the given base side AB.

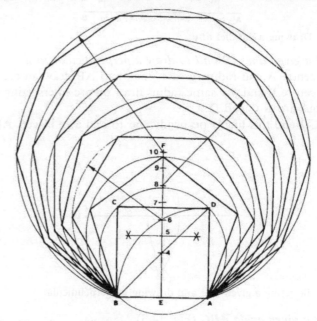

Fig. 14.8 Drawing regular polygons

To inscribe a regular polygon of n sides in a given circle (Fig. 14.9)
1 Divide the diameter AB into *n* equal parts, see Fig. 14.6.
2 With centre A and radius AB draw an arc.
3 With centre B and radius BA draw an arc intersecting the previous arc at C.
4 Join point 2 (always 2) to C and produce to intersect the circle at D.
5 Join AD to construct the first side of the required polygon.
6 Mark off the length of AD around the given circle and join all intersection points to complete the required polygon.

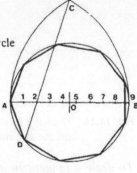

Fig. 14.9 Inscribing a regular polygon in a given circle

14.3 Circles and tangents

A circle is a plane figure bounded by a curve called circumference, which is always the same distance, called radius, from a fixed point called the centre of the circle, see Fig. 14.10.

Diameter is twice the radius and passes through the centre.

An *arc* is the part of the circumference between two points.

A *chord* is a straight line which joins any two points on the circumference. Diameter is the longest chord.

A *segment* is the area which lies between a chord and the arc it intersects.

A *sector* is the area which lies between two radii and the arc between them.

A *quadrant* is the sector between two perpendicular radii.

A *semicircle* is the area which lies between a diameter and the arc it subtends. It is equivalent to two quadrants.

A *tangent* is a straight line which touches the circle at the point of tangency without intersecting it. At the point of tangency any radius forms 90° angle with a tangent.

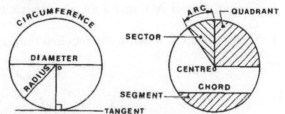

Fig. 14.10 Circles and tangents

To inscribe a circle in a given triangle ABC (Fig. 14.11)

1 Bisect any two angles with bisectors intersecting at O.
2 From O draw OD perpendicular to one of the sides.
3 With centre O and radius equal to OD draw the required circle tangential to each side as shown.

To join two given circles O_1 and O_2 with two fillet arcs of radius R_3 and R_4 (Fig. 14.12)

1 With centre O_1 and radius $R_1 + R_3$ draw an arc.

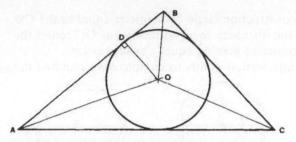

Fig. 14.11 Inscribing a circle in a given triangle

2 With centre O_2 and radius $R_2 + R_3$ draw an arc to intersect the previous arc at A. Join A with O_1 and O_2 to intersect two circles at C and D.
3 With centre A and fillet radius R_3 draw the required fillet of radius R_3 between C and D.
4 With centre O_1 and radius $R_4 - R_1$ draw an arc.
5 With centre O_2 and radius $R_4 - R_2$ draw an arc to intersect the previous arc at B. Join B with O_1 and O_2 and produce to intersect circles at E and F.
6 With centre B and fillet radius R_4 draw the required fillet of radius R_4 between E and F.

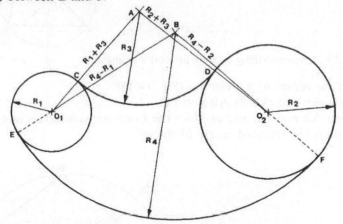

Fig. 14.12 Joining two circles with two fillet arcs of given radii

To draw a hexagon with given distances

(a) across flats, AB in Fig. 14.13a and (b) across corners, CD in Fig. 14.13b. For (b) start construction at step 4.

1 Draw the required distance across flats AB and bisect it at O.
2 Draw two perpendiculars at A and B.
3 Draw a line through O and at 30° to AB to intersect the perpendicular at C and D. CD is the required distance across corners.

4 Draw a construction circle of diameter equal to the CD.
5 Step off the distances equal to the radius OC round the circumference to give six equally spaced points.
6 Join all intersection points to complete the required hexagon.

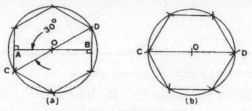

Fig. 14.13 Drawing a hexagon using given distances

To circumscribe a circle around triangle ABC (Fig. 14.14)
1 Bisect any two sides with bisectors intersecting at O.
2 With centre O and radius OA draw the required circle through the points ABC.

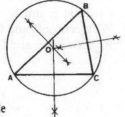

Fig. 14.14 Circumscribing a circle around a triangle

To find the centre of a given arc (Fig. 14.15)
1 Draw any two chords AB and BC.
2 Bisect AB and BC and produce the bisectors to intersect at O which is the required centre of the arc.

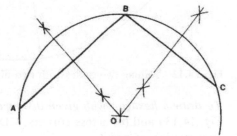

Fig. 14.15 Finding the centre of a given arc

To draw a tangent to a given point P on the circumference of a circle (Fig. 14.16)
1 Join PO and produce to PB so that PB is equal to any distance AP.
2 Draw the perpendicular bisector CD which is the required tangent.

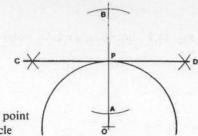

Fig. 14.16 Drawing a tangent to a given point on the circumference of a circle

To draw a tangent to a circle from a given point P outside the circle
1 Join P to the centre of the circle O.
2 Bisect OP at A and with centre A and radius AO draw a semicircle intersecting the circle at B.
3 Draw a line BP which is the required tangent.

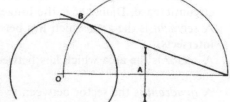

Fig. 14.17 Drawing a tangent to a circle from a given point outside the circle

To draw an internal tangent to two equal circles (Fig. 14.18)
1 Join the centres of the given circles and bisect O_1O_2 at A.
2 Bisect AO_2 at B and with centre B draw a semicircle of radius BO_2 intersecting the circle O_2 at C.
3 Similarly bisect AO_1 and draw a semicircle to intersect the circle O_1 at E.
4 Draw a line CE which is the required tangent.

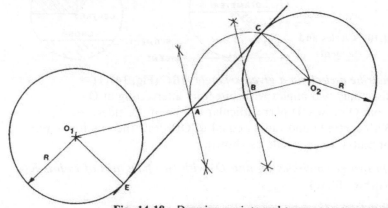

Fig. 14.18 Drawing an internal tangent to two equal circles

128

To draw an external tangent to two circles (Fig. 14.19)

1 Join the centres of the given circles and bisect at A.
2 With centre A draw a semicircle of radius AO_1.
3 With centre O_1 draw an arc of radius $R_1 - R_2$ intersecting the semicircle at B.
4 Join O_1B and produce to intersect the larger circle at C.
5 Join BO_2 and with centre C draw an arc of radius BO_2 intersecting the small circle at D.
6 Draw a line CD which is the required tangent.

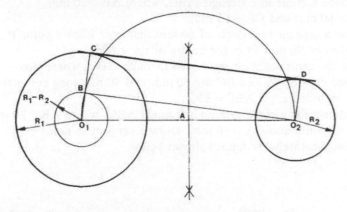

Fig. 14.19 Drawing an external tangent to two circles

To draw an internal tangent to two circles (Fig. 14.20)

1 Join the centres of the given circles and bisect at A.
2 With centre A draw a semicircle of radius AO_1.
3 With centre O_1 draw an arc of radius $R_1 + R_2$ intersecting the semicircle at B. Join B to O_1 intersecting the large circle at C.

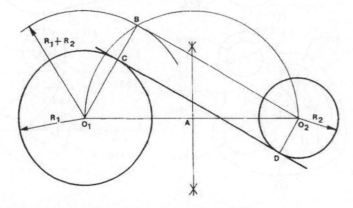

Fig. 14.20 Drawing an internal tangent to two circles

4 Join BO_2 and with centre C draw an arc of radius BO_2 intersecting the small circle at D.
5 Draw a line CD which is the required tangent.

14.4 Construction of ellipses

To draw an ellipse using two concentric circles (Fig. 14.21)

1 Draw two concentric circles equal in diameter to the major axis XX and the minor axis YY of the required ellipse.
   Divide the circles into a number of parts with radial lines crossing the inner and outer circles.
2 Where the radial lines cut the inner and outer circles, draw horizontal and vertical lines respectively. The points of intersection A, B, C, and D are points on the ellipse.
3 Draw a uniform bold curve through the intersection points to form the required ellipse.

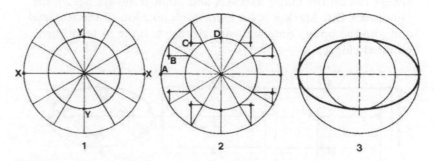

Fig. 14.21 Drawing an ellipse using two concentric circles

To draw an approximate ellipse using instruments (Fig. 14.22)

1 Draw a construction rectangle with XX equal to the major axis and YY equal to the minor axis of the required ellipse.
   Bisect the angle A and the angle B, the bisectors intersecting at C.
2 With the centre at C, draw a circle tangential to XX and YY.
   From A, draw a tangent to the circle with centre C, intersecting YY at D and XX at E.
   By means of dividers or compasses and with the centre at B, using symmetry transfer E to E' and D to D'.
3 With centres at E and E', draw tangential arcs of radius R', and with centres at D and D' draw the remaining arcs of radius R''.
   (The points D and D' may be outside the construction rectangle, depending on the ratio of axes).

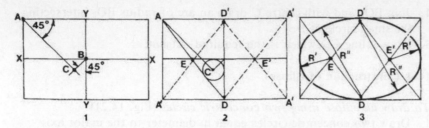

Fig. 14.22 Drawing an approximate ellipse using instruments

To draw an ellipse using the trammel method (Fig. 14.23)

1 Draw the major axis AA and the minor axis BB. Mark off on the trammel (a strip of paper) half the length of the minor axis (EA).

2 Mark off on the trammel half the length of the major axis (EB), measured from the same end E.

3 Position the trammel on the drawing, making sure that point A always lies on the major axis AA and point B always lies on the minor axis BB. Mark a point E for each position of the trammel and join the points obtained with a smooth curve to form the required ellipse.

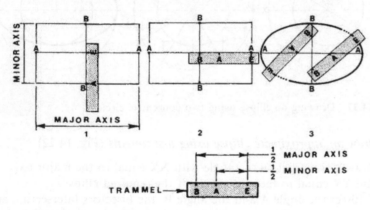

Fig. 14.23 Drawing an ellipse using the trammel method

14.5 Test questions

1 Bisect a 107 mm long line and divide one of the halves into seven equal parts.

2 Draw a heptagon on a 40 mm long side using two different methods.

3 Redraw Fig. 14.5 p. 125, using $R_1 = 30$ mm and $R_2 = 18$ mm measurements.

4 Draw two ellipses of 90 mm major and 50 mm minor axes using an accurate and then an approximate method.

5 Draw an internal tangent to two 50 mm diameter circles 90 mm apart.

6 Draw separately (a) an external and (b) an internal tangent to two circles of 50 mm and 30 mm diameters and 80 mm apart.

7 Redraw Fig. 14.12, p. 127, using $R_1 = 15$ mm, $R_2 = 25$ mm, $R_3 = 35$ mm, $R_4 = 70$ mm and distance between centres $O_1O_2 = 80$ mm.

8 Inscribe a circle in a triangle ABC, where AB = 70 mm, BC = 60 mm and CA = 95 mm.

9 Draw a tangent to a circle of 60 mm diameter from a point P positioned 90 mm from the centre of the circle.

10 Find the centre of a circle to pass through ABC, where two joined chords AB and BC are 50 mm and 40 mm long respectively and included angle ABC = 90°.

11 Circumscribe a circle around a triangle ABC, where AB = 70 mm, BC = 50 mm and CA = 60 mm. Draw a tangent at point C.

12 Draw accurately the figures shown below.

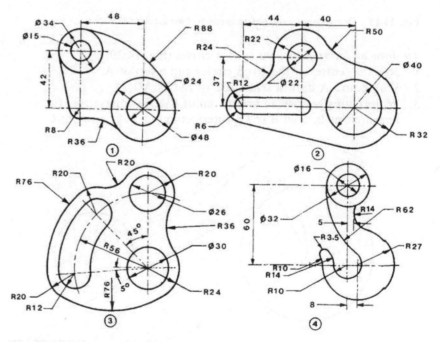

Fig. 14.24 Test question 12

130

15 Loci

The locus of a point is the path traced by the point when it moves in accordance with specified conditions. The plural of locus is *loci*.

15.1 Common loci

Circle

If a point P moves in one plane so that its distance from a fixed point O is constant, then its locus is a circle.

To draw a circle, compasses are set to the required constant distance. With the point of the compass at O (Fig. 15.1), the compass lead then traces out the required circle through P_1, P_2, P_3, etc., where $OP_1 = OP_2 = OP_3 = R$, the radius of the circle.

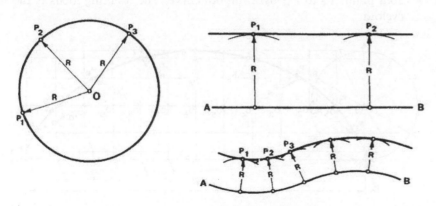

Fig. 15.1 Circle   **Fig. 15.2** Parallel line

Parallel line

If a point P moves in one plane so that its perpendicular distance from a fixed line AB is constant, then its locus is a line parallel to AB.

To draw a parallel line (Fig. 15.2):
1 With centres on AB, strike a number of arcs of radius R equal to the required distance between AB and the parallel line.
2 Draw a common tangent to all these arcs. This is the required parallel line.

Perpendicular line

If a point P moves in one plane so that it is equidistant from two fixed points A and B, then its locus is a straight line perpendicular to AB.

To draw a perpendicular line (Fig. 15.3):
1 With centres at A and B, strike two arcs each of an arbitrary radius R_1 to intersect at P_1 on either side of AB.
2 With the same centres, strike further pairs of arcs with radii R_2, R_3, ..., etc. and intersection points P_2, P_3,... on either side of AB. A straight line drawn through P_1, P_2, P_3,..., etc. is the required line perpendicular to AB (and is also the bisector of AB).

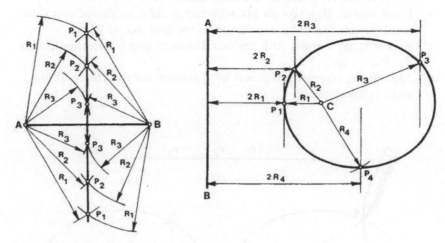

Fig. 15.3 Perpendicular line   **Fig. 15.4** Ellipse

Ellipse – method 1

If a point P moves in one plane so that its distance from a fixed point C and its perpendicular distance from a fixed line AB are always in the same ratio $1:n$, where n is any number greater than 1, then the locus of the point is an ellipse.

To draw an ellipse (Fig. 15.4):
1 Taking a distance ratio of, say, 1:2, draw a line parallel to AB and at an arbitrary distance $2R_1$ from it.
2 From centre C, strike an arc of radius R_1 to intersect this line at P_1.
3 Repeat steps 1 and 2 with radii R_2, R_3, ..., etc. to give intersection points P_2, P_3, ..., etc.
4 Join all the intersection points by a smooth curve. This curve is the required ellipse.

131

Ellipse – method 2

If a point P moves in one plane so that the sum of its distances from two fixed points A and B is constant, then its locus is again an ellipse.

If a piece of string of total length equal to AP + PB is fixed with its ends at A and B and is kept taut by a pencil held against it inside the loop so formed, moving the pencil will produce a locus which is an ellipse.

To draw an ellipse (Fig. 15.5):

1. For measuring purposes, draw a construction line of total length equal to AP + PB (Fig. 15.5(a)).
2. Draw the two fixed points A and B (Fig. 15.5(b)).
3. From centre A, strike an arc with radius AP_1.
4. From centre, B, strike an arc with radius BP_1, measured from the construction line. This arc intersects the first one at P_1.
5. Repeat steps 3 and 4 with various distances and intersection points P_2, P_3, ..., etc.
6. Join all the intersection points by a smooth curve. The resulting locus is an ellipse.

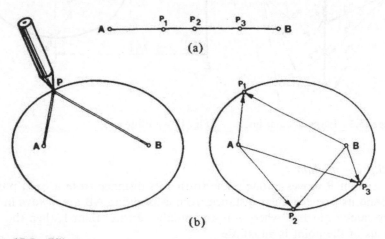

Fig. 15.5 Ellipse

Cycloid

A cycloid is the locus of a point on the circumference of a circle which rolls without slip along a fixed straight line.

When a disc with a pencil positioned on its periphery is rolled along a straight edge without slip, the moving pencil will trace the cycloid curve.

The profile of certain gear teeth is based on the cycloid curve.

To draw a cycloid:

1. Draw a circle of radius R, say, with centre O and mark off, say, twelve points equally spaced around its circumference as shown in Fig. 15.6. Number the points in a clockwise direction, the bottom point being numbered twice (0 and 12).
2. Through each point draw a horizontal construction line.
3. On the construction line passing through O, mark out a distance equal to the circumference of the circle ($2\pi R$) and divide it into twelve equal parts with points O, O_1, ..., O_{12}.
4. From centre O, strike an arc of radius R to intersect the construction line from O on the circumference. Call the intersection point P_0.
5. From centre O_1, strike another arc of radius R to give point P_1 where it intersects the construction line from 1 on the circumference.
6. From centres O_2, O_3, ..., O_{12}, strike arcs in the same way to give points P_2, P_3, ..., P_{12}.
7. Join points P_0 to P_{12} by a smooth curve. The resulting locus is the cycloid.

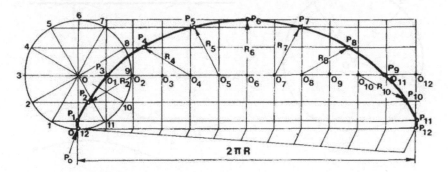

Fig. 15.6 Cycloid

Epicycloid and hypocycloid

An *epicycloid* is the locus of a point on the circumference of a circle which rolls without slip along the *outside* of a fixed base circle.

A *hypocycloid* is the locus of a point on the circumference of a circle which rolls without slip along the *inside* of a fixed base circle.

The profile of certain gear teeth is based on a combination of epicycloid and hypocycloid curves, as shown in Fig. 15.7(b).

To draw an epicycloid or hypocycloid (Fig. 15.7(a)):

1. Strike an arc of the appropriate radius for the fixed base circle and draw the rolling circle of radius R touching it – above the arc for the epicycloid, below it for the hypocycloid.

132

2 Mark off, say, twelve equally spaced points around the circumference of the rolling circle and number them as shown – clockwise for the epicycloid, anticlockwise for the hypocycloid.

3 With centres at the centre of the fixed base circle, draw construction arcs through these points and through the centre of the circle, O.

4 Along the arc of the fixed base circle, mark off a length equal to the circumference of the rolling circle ($2\pi R$). This length subtends an angle $\theta = R/R_b \times 360°$ at the centre of the base-circle arc.

5 Divide this angle θ into twelve equal parts by lines passing through the construction arcs. On the construction arc passing through the centre of the rolling circle, O, number the intersection points O_1, O_2,..., O_{12} as shown.

6 From centre O_1, strike an arc of radius R (rolling-circle radius) to give point P_1 where it intersects the construction arc passing through point 1 on the rolling circle.

7 From centre O_2, strike another arc of radius R to give point P_2 where it intersects the construction arc passing through point 2 on the rolling circle.

8 From points O, O_3, O_4, ..., O_{12}, strike arcs in the same way to give points P_0, P_3, P_4, ..., P_{12}.

9 Join points P_0 to P_{12} by a smooth curve. The resulting locus will be the epicycloid or hypocycloid as appropriate.

Involute

An involute is the locus of a point on a straight line which rolls without slip around a circle.

If a length of string with a pencil attached to its free end is unwound under tension from around the circumference of a fixed disc, then the moving pencil will trace an involute curve.

Part of the involute curve is used for gear-tooth profiles, as shown in Fig. 15.8(b); it has many advantages over the cycloid curves.

To draw an involute:

1 Draw the base circle and mark off, say, twelve equally spaced points around its circumference, numbering the points as shown in Fig. 15.8(a).

2 At point 1, draw a tangent of length equal to one twelfth of the circumference of the circle. The 'free' end of the tangent is point P_1.

3 At point 2, draw a tangent of length equal to two twelfths of the circumference of the circle, to give point P_2.

4 In the same way, from points 3, 4,..., 12 draw tangents of length $3/12$, $4/12$, ..., $12/12$ of the circumference of the circle respectively, to give points P_3, P_4, ..., P_{12}.

5 With the aid of a French curve, join points P_1 to P_{12} by a smooth curve. The resulting locus will be the involute.

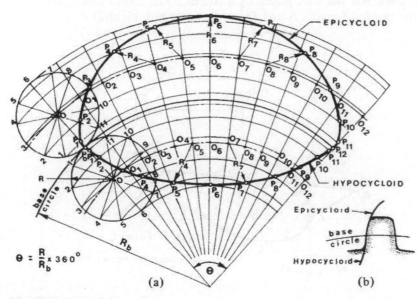

Fig. 15.7 Epicycloid and hypocycloid

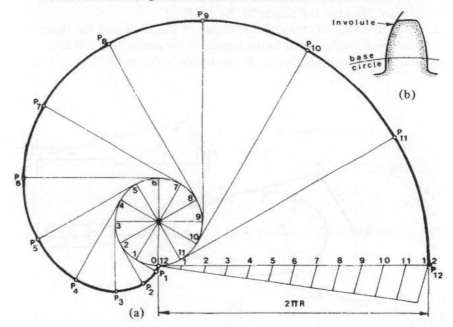

Fig. 15.8 Involute

15.2 Mechanisms

A mechanism is a machine or part of a machine consisting of a system of moving parts. Any rigid piece of the mechanism pivoted at the ends is called a *link*. Arrangements of links, pivots, and slides are often used to convert circular motion into reciprocating or oscillating motion, and vice versa.

When designing a mechanism to perform a specific task, the loci of various points on the mechanism must be drawn to determine the velocities, accelerations, and forces involved in the motion produced. It is also necessary to study the movements of the links so that clearances may be checked and safety guards or shields may be designed to protect the users of the machine.

Two very common types of mechanism are the slider-crank and the four-bar chain.

Slider-crank

The slider-crank is a simple mechanism in which rotary motion is converted to linear motion, or vice versa. Figure 15.9(a) shows a crank AB which rotates about A and is joined by the connecting rod BC to the piston (slider) C which slides along the axis AC.

To plot the locus of a point P on the mechanism (Fig. 15.9(b)):

1 Draw a circle of radius equal to the crank length AB and mark off, say, twelve points equally spaced on its circumference as shown. Number the points $B_1, B_2, ..., B_{12}$.

2 Draw the axis AC through the centre of the circle and the slider and, with radius equal to the length of the connecting rod BC, strike arcs from $B_1, B_2, ..., B_{12}$ to intersect AC at $C_1, C_2, ..., C_{12}$.

3 Join $B_1C_1, B_2C_2, ..., B_{12}C_{12}$ to give the positions of the connecting rod BC for twelve different positions of the crank AB.

4 With the same centres, $B_1, B_2, ..., B_{12}$ and radius BP, strike arcs to intersect $B_1C_1, B_2C_2, ..., B_{12}C_{12}$ at $P_1, P_2, ..., P_{12}$.

5 Join points $P_1, P_2, ..., P_{12}$ by a smooth curve. This is the locus of point P.

Alternatively, a trammel method may be used. The trammel, which may be a piece of paper with a straight edge on which the relevant points are marked, represents the connecting rod and its movements are plotted. The trammel method enables a large number of points to be obtained quickly, and so is widely used by designers in industry.

To plot the locus of a point P on the mechanism using a trammel:

1 On an axis AC with centre A, draw a circle of radius equal to the crank length AB.

2 Mark points BPC on the edge of the paper to represent the connecting rod.

3 Place the paper in various positions such that point B always touches the circumference of the circle and point C is always on the axis AC. For each position, mark the position of P.

4 Join the successive positions of P to obtain the required locus.

Four-bar chain

The four-bar chain is a simple mechanism which consists of two cranks, AB and CD, joined by a rod BC. The fourth link is between the two fixed pivots A and D, as shown in Fig. 15.10(a).

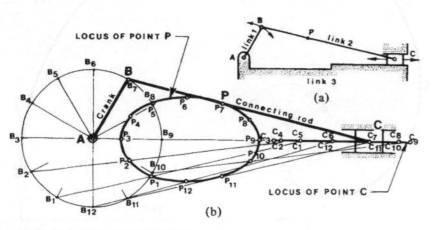

Fig. 15.9 Slider-crank mechanism

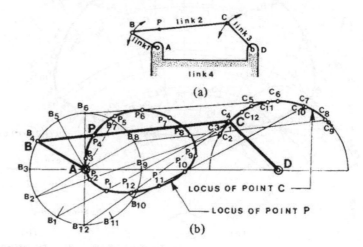

Fig. 15.10 Four-bar-chain mechanism

To plot the locus of a point P (Fig. 15.10(b)):

1 Draw a circle of radius AB with centre A and another of radius DC with centre D.
2 Mark off, say, twelve equally spaced points on the circle of radius AB to correspond to twelve different positions of the crank AB. Number the points B_1, B_2, ..., B_{12} as shown.
3 With B_1, B_2, ..., B_{12} as centres, strike arcs of radius equal to the length of the connecting rod BC to intersect the circumference of the circle of radius DC at points C_1, C_2, ..., C_{12}.
4 Join B_1C_1, B_2C_2, ..., $B_{12}C_{12}$ to give the positions of the connecting rod BC for twelve different positions of the crank AB.
5 Again with centres, B_1, B_2, ..., B_{12}, strike arcs of radius BP to intersect B_1C_1, B_2C_2, ..., $B_{12}C_{12}$ at P_1, P_2, ..., P_{12}.
6 Join points P_1 to P_{12} by a smooth curve. This is the locus of point P. Alternatively, a trammel method may be used.

Sliding link

If a ladder AB is propped against a wall and the bottom end slides outwards from B_1 to B_2, then the top of the ladder will slip correspondingly from A_1 to A_2, since the length of the ladder remains constant.

A sliding link is one which is free to move in one plane such that one end of the link is always in contact with a line OY and the other end is always in contact with another line OX.

To plot the locus of a point P on the link (Fig. 15.11):

1 With centres A_1, A_2, ..., etc. on line OY and radius equal to the length of the link AB, strike a number of arcs to intersect OX at B_1, B_2, ..., etc.
2 Join A_1B_1, A_2B_2, ..., etc. to obtain various positions of the link.
3 Again with centres A_1, A_2, ..., etc., strike arcs of radius AP to intersect A_1B_1, A_2B_2, ..., etc., at P_1, P_2, ..., etc.
4 Join the points P_1, P_2, ..., etc. with a smooth curve. This is the required locus.

Quick-return mechanism

The quick-return mechanism shown in Fig. 15.12 is used to reduce the time wasted during the non-cutting return stroke of a shaping machine, etc.

A pinion drives a gear wheel with a uniform angular velocity. Attached to the gear wheel is a pin which rotates with it and at the same time moves a sliding block up and down inside the slotted link, which in turn oscillates about its pivot.

The shaping-machine ram, which is joined to the top of the slotted link by means of a connecting link, makes its cutting stroke while the pin is travelling through the larger arc, of about 240°. The non-cutting return stroke is made while the pin travels through the smaller arc, of about 120°. Thus the ram moves back twice as fast as it moves forward on its cutting stroke.

The length of the ram stroke can be adjusted by changing the radial distance of the pin from the centre of the gear wheel.

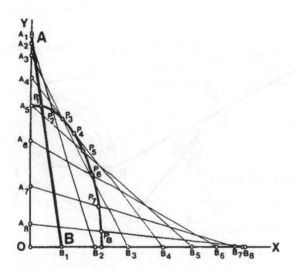

Fig. 15.11 Sliding link

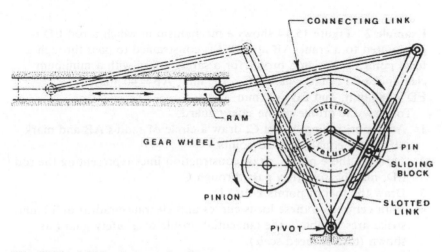

Fig. 15.12 Quick-return mechanism

135

15.3 Worked examples

Example 1 If a point P moves in one plane so that its distances from two fixed points A and B are always on the same ratio 2:1, then its locus is a circle.

To draw a circle where A and B are 60 mm apart (Fig. 15.13):

1 Strike an arc with radius R_1 of 20 mm from point B and another arc with radius $2R_1$ from point A to intersect the first arc at P_1.
2 Repeat the procedure with different radii, still in the ratio 2:1, to give intersection points P_2, P_3, ..., etc.
3 Join all the intersection points with a smooth curve. The resulting locus is the circumference of a circle as shown (to a reduced scale).

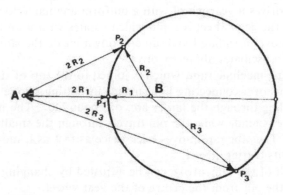

Fig. 15.13 Circle

Example 2 Figure 15.14 shows a mechanism in which a rod ED is pin-jointed to a crank AB at B and is constrained to pass through a fixed point C. Design a profile for a safety guard with a minimum clearance of 12 mm, where AC = 100 mm, AB = 40 mm, ED = 180 mm, and EB = 20 mm.

To draw the profile of the safety guard:

1 After positioning A and C, draw a circle of radius AB and mark off twelve equally spaced points.
2 Through these points, draw construction lines representing the rod ED, making sure they pass through C.
3 Draw the loci of points E and D.
4 With centres on these locus curves and clearance radius of 12 mm, strike arcs to obtain the tangential profile of a safety guard as shown (to a reduced scale).

Alternatively, a trammel method may be used.

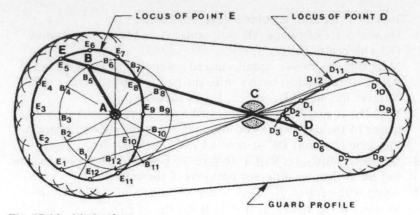

Fig. 15.14 Mechanism

Example 3 Figure 15.15 shows a wrapping-machine linkage pivoted at D and A and pin-jointed at C, B, and E. Link EP, length 100 mm, is constrained to pass through the point F, and link AB, length 50 mm, oscillates about A between B_1 and B_2. Design an outline of a safety guard with a minimum clearance of 10 mm.

To draw the outline of a safety guard:

1 Draw the locus of point E (Fig. 15.10). Using this locus, draw the required locus of point P.
2 Draw the outline of a safety guard as shown (to a reduced scale).

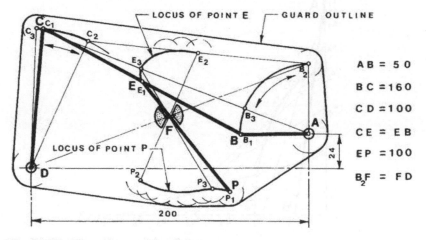

Fig. 15.15 Wrapping-machine linkage

AB = 50
BC = 160
CD = 100
CE = EB
EP = 100
B_2F = FD

15.4 Test questions

1 Define the locus of a moving point.
2 Relate the locus definition to the circle, ellipse, and cycloid.
3 Define the epicycloid, hypocycloid, and involute and explain for what purposes these curves are used in engineering.
4 Explain the importance for engineering designers of determining the locus of a point on a moving mechanism.
5 Construct a perpendicular line to bisect a line AB 47 mm long.
6 Draw a line AB and a point C 60 mm from it. Then draw the locus of point P which moves in such a way that it is always at one chosen distance from the point C and at double the chosen distance measured perpendicularly from the line AB.
7 Draw the locus of a point P which moves so that the sum of its distances AP + PB from two fixed points A and B remains constant at 120 mm. The distance between A and B is 75 mm.
8 Draw a cycloid of a rolling circle 80 mm diameter.
9 Draw the epicycloid and hypocycloid for a rolling-circle radius of 30 mm and a base-circle radius of 90 mm.
10 (a) Define the involute and explain why this curve is important in engineering design.
   (b) Draw the involute on a base circle of 50 mm diameter.
11 Draw the locus of point P for the slider-crank mechanism shown in Fig. 15.16.
12 Draw the locus of point P for the four-bar chain shown in Fig. 15.17.
13 Figure 15.18 shows a pin-jointed mechanism. The cranks AB and CD revolve about A and C at the same speed.
   Draw the locus of the points E and F and then the outline of a safety guard with a minimum clearance of 10 mm.
14 Draw the machine mechanism shown in Fig. 15.19 and plot the locus of point E for a complete revolution of the crank AB.
   Draw also the guard outline with a minimum clearance of 12 mm.
15 A mechanism used in a textile machine is shown in Fig. 15.20. Crank AC rotates about A. Connecting rod BE is pin-jointed at B, with E constrained to reciprocate vertically. DF is pin-jointed at D but is free to slide through the swivelling guide C at all times.
   Draw the loci of points D and F and construct the profile of a suitable guard to enclose the mechanism with a minimum clearance of 15 mm.
16 Draw the locus of point E of the mechanism shown in Fig. 15.21, where BE is constrained to pass through the tee-piece at C.
17 Figure 15.22 shows a mechanism pivoted at A and D and pin-jointed at C, E, and B. The link EG is free to slide through the swivelling guide F at all times. Plot the path of points E and G and construct a safety guard with a minimum clearance of 16 mm.

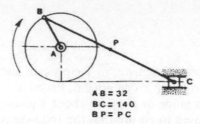

AB = 32
BC = 140
BP = PC

Fig. 15.16   Test question 11

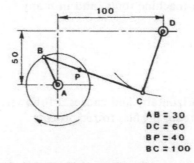

AB = 30
DC = 60
BP = 40
BC = 100

Fig. 15.17   Test question 12

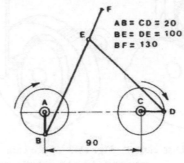

AB = CD = 20
BE = DE = 100
BF = 130

Fig. 15.18   Test question 13

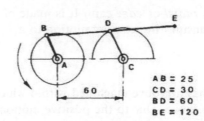

AB = 25
CD = 30
BD = 60
BE = 120

Fig. 15.19   Test question 14

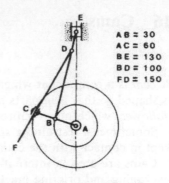

AB = 30
AC = 60
BE = 130
BD = 100
FD = 150

Fig. 15.20   Test question 15

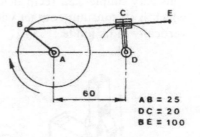

AB = 25
DC = 20
BE = 100

Fig. 15.21   Test question 16

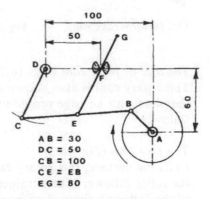

AB = 30
DC = 50
CB = 100
CE = EB
EG = 80

Fig. 15.22   Test question 17

16 Cams

A *cam* is a component which may rotate, oscillate, or reciprocate and is shaped so that it imparts motion to another component, called a *follower*, which may reciprocate in a guide or oscillate about a pivot.

Sometimes a restraining spring is used to ensure that the follower is kept in contact with the cam.

Cams are used in internal-combustion engines to operate valves, in packaging and printing machinery, in machine tools, and in many other industrial applications.

16.1 Main types of cam

The wedge cam (Fig. 16.1)
This very simple cam reciprocates horizontally and causes a follower, which is in constant contact with the cam profile, to reciprocate vertically in its guide.

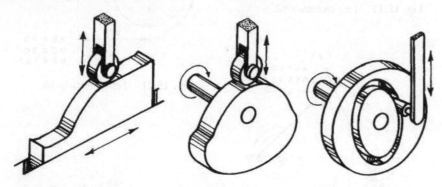

Fig. 16.1 Wedge cam   **Fig. 16.2** Disc cam   **Fig. 16.3** Face cam

The disc or plate cam (Fig. 16.2)
This rotary cam is also known as a *radial* or *edge* cam. It is made of a flat plate with an edge profile to transmit the required motion to a follower.

The face cam (Fig. 16.3)
In its flat surface, this rotary cam has a groove machined within which the roller follower is constrained to move. Due to the positive motion of the follower, there is no need for a restraining spring.

The cylindrical cam (Fig. 16.4)
In its curved surface, this rotary cam has a groove machined within which the roller follower is constrained to move. The reciprocating positive motion of the follower is parallel to the cam axis.

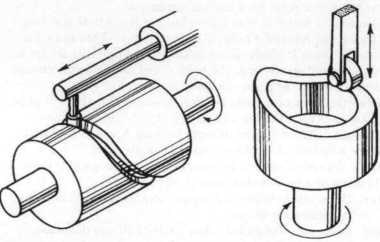

Fig. 16.4 Cylindrical cam   **Fig. 16.5** End cam

The end cam (Fig. 16.5)
This cylindrical cam has its end machined to the required shape.

16.2 Main types of cam follower
There are three main types of follower: knife-edge, flat and roller.

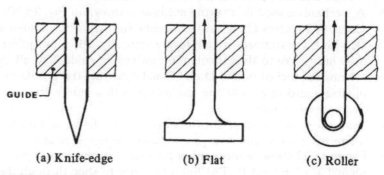

(a) Knife-edge   (b) Flat   (c) Roller

Fig. 16.6 Cam followers

The knife–edge follower (Fig. 16.6(a))

This has the advantage that it can follow any complicated cam profile. However, it is not often used as it wears rapidly, due to high pressures and sliding friction.

The flat follower (Fig. 16.6(b))

This cannot be used for concave cam profiles. It wears slower than a knife-edge follower, since the points of contact move across the surface of the follower according to the changing profile of the cam.

The roller follower (Fig. 16.6(c))

This has the advantage that wear is minimised, due to rolling rather than sliding friction. The cam profile must not incorporate any concave forms with a radius smaller than the radius of the roller.

The cam follower may have a reciprocating or oscillating motion and may be positioned in line with the cam centre line or be offset, as shown in Fig 16.7.

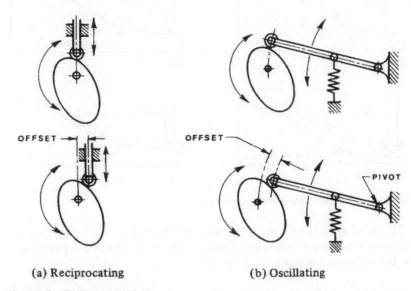

(a) Reciprocating   (b) Oscillating

Fig. 16.7 Follower motion

16.3 Types of follower motion

To design the profile of a rotary cam in order to impart a required motion to a follower, it is very convenient to draw first a displacement diagram. On this diagram, the linear displacement of the follower is plotted vertically against angular displacement for a complete rotation of the cam.

There are three fundamental types of motion which may be imparted to followers: uniform velocity, uniform acceleration and retardation, and simple harmonic motion.

Uniform velocity (Fig. 16.8)

The displacement diagram will be a straight sloping line, as the displacement is proportional to the angle turned through, and equal angles turned through by the cam will produce equal increments of rise or fall of the follower.

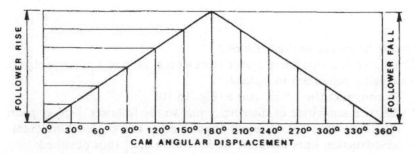

Fig. 16.8 Uniform velocity

Uniform acceleration and retardation

The uniform acceleration and retardation curves are parabolic in form. To construct them (Fig. 16.9).

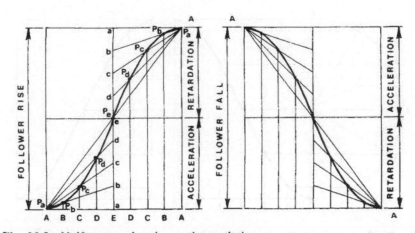

Fig. 16.9 Uniform acceleration and retardation

1 Divide the rise of the follower and the angular displacement of the cam into the *same* number of equal parts by points A, B, ..., etc. and a, b, ..., etc. as shown.

2 For the acceleration curve, draw construction lines radiating from the *initial* point A to a, b, ..., etc, and for the retardation draw construction lines radiating from the *final* point A to a, b, ..., etc.

3 Mark point P_b where the line Ab intersects the vertical through B, point P_b, P_c, ..., etc. will be the required acceleration and retardation curves.

Simple harmonic motion (s.h.m.)

This curve is a sine curve and it represents a motion similar to that of a swinging pendulum in a clock.

To construct the s.h.m. curve (Fig. 16.10):

1 Draw a semicircle of diameter equal to the follower displacement. Divide it into, say, six equal angular parts and project horizontal construction lines from the intersection points thus obtained.

2 Divide the cam displacement into the same number of equal parts (six) and project verticals.

3 Join the corresponding points of intersection of the vertical and horizontal lines to obtain the smooth s.h.m. curve.

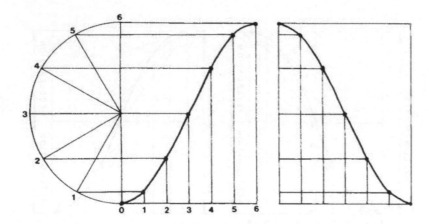

Fig. 16.10 Simple harmonic motion (s.h.m.)

16.4 Construction of cam profiles

1 Figure 16.11(a) shows a follower displacement diagram (F.D.D.) for a wedge cam.

2 Figure 16.11(b) shows that the profile of a wedge cam is identical to the follower displacement diagram when a knife-edge follower is used.

3 Figure 16.11(c) shows that the follower displacement diagram is identical to the locus of the centre of a roller follower. The cam profile is drawn tangential to construction circles representing the roller in various positions.

4 Figure 16.11(d) shows that the follower displacement diagram is identical to the locus of the central point on the contact surface of a flat follower. The cam profile is drawn to touch the points of contact of construction lines representing the flat follower in various positions.

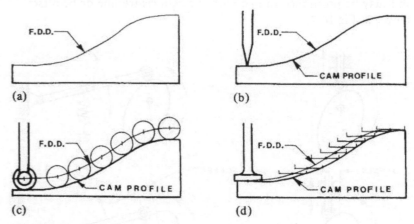

Fig. 16.11 Various shapes of wedge cam

It can be seen from Fig. 16.11 that the profiles of wedge cams may be of varying shapes when transmitting identical motion, depending on the different types of follower used.

Example 1 – knife-edge follower

Cam specification: disc cam, minimum radius 40 mm; shaft diameter 35 mm; rotation anticlockwise.

Displacement and motion:

   $0°-90°$, rise of 30 mm with uniform acceleration
  $90°-180°$, rise of 30 mm with uniform retardation
$180°-240°$, dwell (no motion or rest)
$240°-360°$, fall with uniform velocity

To draw the cam profile (Fig. 16.12):

1 Draw the follower displacement diagram, incorporating the cam minimum radius, which is the nearest approach of the follower to the centre of the cam.
2 Draw two circles, one of 40 mm cam minimum radius and the other of 100 mm cam maximum radius (i.e. the minimum cam radius plus the maximum follower displacement).
3 Draw radial lines at 30° intervals, numbering them in the reverse direction to the cam rotation. If extra accuracy is required, 15° or smaller intervals may be used as shown in the 300°–360° interval. *Note*: for construction purposes, the cam may be considered to be stationary while the follower rotates in the reverse direction about the cam.
4 Transfer all vertical distances from the follower displacement diagram to the corresponding radial lines, measuring from the centre.
5 With the help of a French curve, draw a smooth curve through the points obtained, to give the required cam profile.

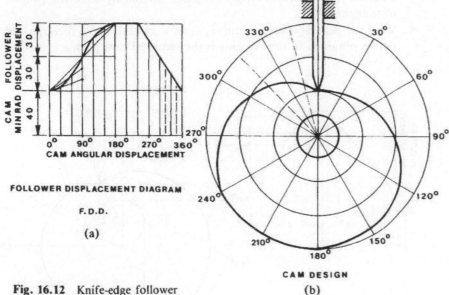

FOLLOWER DISPLACEMENT DIAGRAM

F.D.D.

(a)

CAM DESIGN

(b)

Fig. 16.12   Knife-edge follower

Example 2 – roller follower
Cam specification: as example 1 (minimum radius is 40 mm minus the roller radius; i.e. 40 – 6 = 34 mm).
Displacement and motion : as example 1.
Follower: 12 mm diameter roller.

To draw the cam profile (Fig. 16.13):

1 Draw the locus of the roller centre, which is an identical curve to the cam profile already constructed for the knife-edge follower in example 1 (Fig. 16.12(b)).
2 With centres on this locus and radius 6 mm, draw a number of construction circles representing the roller in various positions.
3 Draw the best tangential curve to these circles to give the required cam profile.

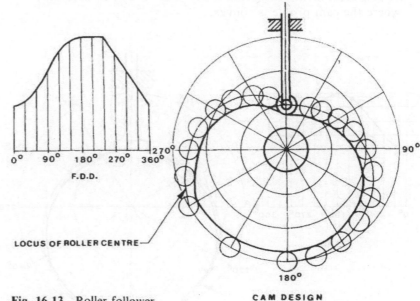

F.D.D.

LOCUS OF ROLLER CENTRE

CAM DESIGN

Fig. 16.13   Roller follower

Example 3 – flat follower
Cam specification: disc cam, minimum radius 46 mm; shaft diameter 38 mm; rotation clockwise
Displacement and motion:

   0°–180°, rise of 48 mm with s.h.m.
180°–360°, fall of 48 mm with s.h.m.

Follower: flat, with 30 mm long contact surface.

To draw the cam profile (Fig. 16.14):

1 Draw the follower displacement diagram, F.D.D., incorporating the minimum cam radius.
2 Draw two circles, one of 46 mm cam minimum radius and another of 94 mm cam maximum radius.

141

3 Draw radial lines at 30° intervals, labelling them anticlockwise.
4 Transfer all the vertical distances from the follower displacement diagram to the corresponding radial lines of the cam, measuring from the centre.
5 At each of the plotted points on the radial lines, draw a line 30 mm long perpendicular to and bisected by the radial lines to represent the flat contact surface of the follower. These lines will be tangents to the cam profile.
6 Draw the best smooth curve to touch these tangents at the points of contact, remembering that a flat follower can be used only where the cam profile is convex.

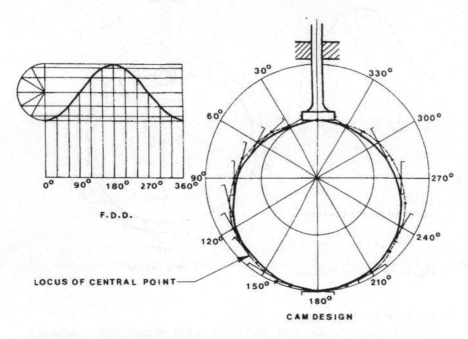

F.D.D.

LOCUS OF CENTRAL POINT

CAM DESIGN

Fig. 16.14 Flat follower

Example 4 – offset knife-edge follower
Cam specification: disc cam, minimum radius 35 mm; rotation anticlockwise.
Displacement and motion:

0°–150°, 54 mm rise with uniform velocity
150°–180°, dwell
180°–360°, 54 mm fall with s.h.m.

Follower: knife-edge, offset 18 mm to the right of the cam centre line.

To draw the cam profile (Fig. 16.15):
1 Draw the base line of the follower displacement diagram, F. D. D., with vertical lines representing 30° intervals of rotation.
2 From point A on the base line, step off the 18 mm offset to obtain the point B. With centre B and radius 35 mm (the cam minimum radius) strike an arc intersecting the nearest vertical line at C.
3 From C draw a horizontal line, from which the follower curves of motion should be constructed in the usual way.
4 To draw the cam, with centre B draw a construction circle of 18 mm radius, equal to the amount of offset, and divide it with radial lines at 30° intervals.
   At the intersection points on the circumference, draw tangents to represent twelve positions of the follower.
5 With centre B draw a circle of 35 mm minimum radius, intersecting the vertical tangent at C.
6 Along this tangent, step off a distance CD equal to the 54 mm maximum follower rise. With centre B, draw a circle through the point D and label all intersection points in a clockwise direction.
7 Transfer all the vertical distances from the follower displacement diagram to the corresponding tangents, measuring from the points of tangency.
8 With the help of a French curve, draw a smooth curve through the points obtained in step 7. This is the required cam profile.

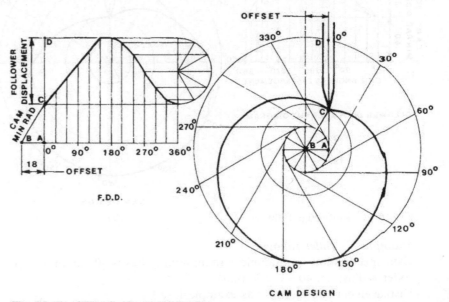

F.D.D.

OFFSET

CAM DESIGN

Fig. 16.15 Offset knife-edge follower

142

Example 5 – offset roller follower

Cam specification: as example 4

Displacement and motion: as example 4

Follower: 16 mm diameter roller offset 18 mm to the right of the cam centre line.

To draw the cam profile (Fig. 16.16)

1 Draw the locus of the roller centre, which is an identical curve to the cam profile already constructed for the knife-edge follower in example 4 (Fig. 16.15).
2 With centres on this locus, draw a number of construction circles of 16 mm diameter to represent the roller in various positions.
3 Draw the best tangential curve to these circles to give the required cam-profile curve.

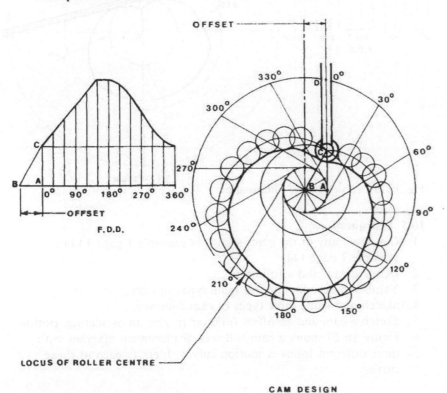

Fig. 16.16 Offset roller follower

Example 6 – face cam

Cam specification: face cam; minimum cam and roller centre distance 28 mm; shaft diameter 20 mm; rotation clockwise.

Displacement and motion:

0°–180°, 36 mm rise with s.h.m.

180°–360°, 36 mm fall with s.h.m.

Follower: 18 mm diameter roller

To draw the cam groove (Fig. 16.17):

1 Draw the follower displacement diagram, F.D.D., incorporating the minimum cam radius.
2 Draw a construction circle of 80 mm cam maximum radius.
3 Draw radial lines at 30° intervals and label them in an anticlockwise direction.
4 Transfer all the vertical distances from the follower displacement diagram to the corresponding radial lines, measuring from the centre.
5 Draw a smooth curve through the points obtained to give the locus of the centre of the follower and, with centres on this locus, draw a number of construction circles of 18 mm diameter to represent the roller follower in various positions.
6 Draw the two best tangential curves touching the construction rollers. These form the required cam groove.

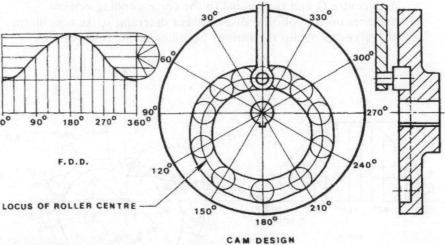

Fig. 16.17 Face cam

Example 7 – radial-arm roller follower

Cam specification: disc cam, minimum radius 20 mm; shaft diameter 22 mm; rotation clockwise.

Displacement and motion:

0°–120°, 36 mm rise with uniform velocity

120°–240°, dwell

240°–360°, 36 mm fall with uniform velocity

143

Follower: radial arm, 110 mm long, with a 16 mm diameter roller follower, positioned on the cam centre line at 110 mm to the left of the cam centre.

To draw the cam profile (Fig. 16.18)

1 Draw the required follower displacement diagram, F.D.D.
2 Position the pivot centre A_1 relative to the cam centre O as shown.
3 With centre O, draw a circle of 110 mm radius through the pivot centre A_1 and draw radial intersection points A_2, A_3, ..., A_{12} at $30°$ intervals representing the pivot at twelve positions of the cam.
4 With centre O, draw a circle of 28 mm (20 mm cam minimum radius plus 8 mm roller radius) representing the lowest position of the roller centre.
5 With centre O draw a circle of 64 mm (28 mm plus the maximum follower rise of 36 mm) representing the highest position of the roller centre.
6 With centre A_1 and radius A_1O (110 mm), draw an arc to intersect the lowest and highest position of the roller-centre circles at B_1 and C_1 respectively.

   Repeat this procedure with centre A_2 to intersect at B_2 and C_2 etc.
7 With centre O and radii equal to the corresponding vertical distances on the follower displacement diagram, strike arcs on the BC curves to obtain the various positions of the roller centres.

8 Draw circles of 16 mm roller diameter at these positions.
9 Draw the best cam-profile curve tangential to these circles.
Note: If the pivot was positioned to the right of the cam and the cam rotation was anticlockwise, the construction would be the mirror image of that in Fig. 16.18, as shown in Fig.16.19.

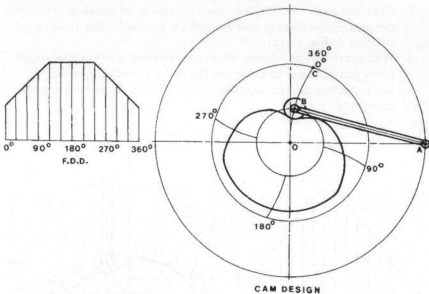

Fig. 16.19 Anticlockwise cam rotation

16.5 Test questions

1 Construct any of the cams shown in example 1 page 140 to example 7 page 144.
2 Define a cam and a follower.
3 Name and describe the five main types of cam.
4 Sketch and name three types of cam follower.
5 Sketch a cam and an offset follower to give an oscillating motion.
6 Figure 16.20 shows a cam-follower displacement diagram with three different follower-motion curves. Identify each of these curves.

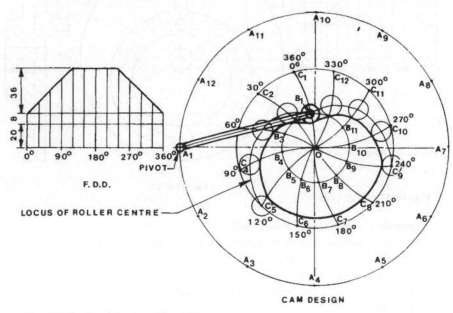

Fig. 16.18 Radial-arm roller follower

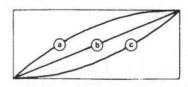

Fig. 16.20 Test question 6

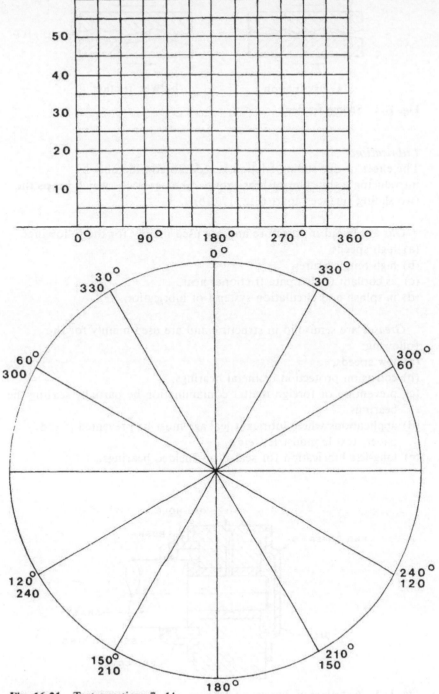

Fig. 16.21 Test questions 7–11

7 Construct three cam-follower displacement diagrams between 0° and 180° to give a rise of 60 mm and the following motions:
(a) uniform velocity,
(b) uniform acceleration and retardation,
(c) simple harmonic motion.
(Tracing paper and Fig. 16.21 may be used).

8 Construct a cam to satisfy the following design specifications: a disc cam of minimum radius 40 mm mounted on a camshaft of 40 mm diameter to rotate in an anticlockwise direction with a knife-edge follower to have the following displacement and motion:

0°–180°, rise of 70 mm with simple harmonic motion;
180°–360°, fall of 70 mm with uniform velocity.

(You may draw to half full-size scale, using tracing paper and Fig. 16.21).

9 Using tracing paper and Fig. 16.21, construct a cam of minimum radius 15 mm to rotate in an anticlockwise direction. The cam-follower displacement and motion is to be a rise of 40 mm with simple harmonic motion for half a revolution and a fall with simple harmonic motion for the remaining half a revolution. Construct and compare superimposed cam profiles for the following followers: (a) knife-edge, (b) flat with 20 mm long contact surface.

10 Construct a cam to satisfy the following design specifications: a disc cam of a minimum radius 40 mm mounted on a camshaft of 24 mm diameter to rotate in a clockwise direction; a cam follower with a roller of 20 mm diameter to have the following displacement and motion:

0°–120°, rise of 30 mm with uniform acceleration;
120°–240°, rise of 30 mm with uniform retardation;
240°–360°, fall of 60 mm with uniform velocity.

(You may draw to half full-size scale, using tracing paper and Fig. 16.21).

11 Construct a cam of minimum radius 30 mm mounted on a camshaft of 30 mm diameter to rotate in an anticlockwise direction; a flat cam follower with 24 mm long contact surface to have the following displacement and motion:

0°–150°, rise of 80 mm with uniform velocity;
150°–180°, dwell;
180°–360°, fall of 80 mm with simple harmonic motion.

(You may draw to half full-size scale, using tracing paper and Fig. 16.21).

17 Bearings

17.1 Reasons for the use of bearings

A bearing is a support provided to hold a component in its correct position while at the same time allowing it to rotate or slide.

There are two main types of bearing to support different loads (forces):

(i) *journal bearings* support radial loads, which act at right angles to the axis of the shaft as shown in Fig. 17.1(a);

(ii) *thrust bearings* support axial loads, which act along the axis of the shaft as shown in Fig. 17.1(b).

Also, bearings may be plain or of the rolling ('antifriction') type.

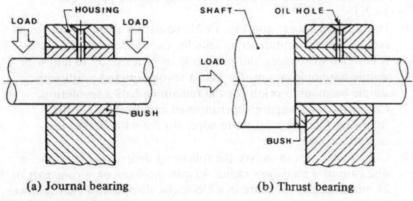

(a) Journal bearing   (b) Thrust bearing

Fig. 17.1 Plain bearings

Friction

Resistance to sliding of two surfaces in contact is called friction and is caused mainly by very small imperfections on the surfaces, even if machined. When magnified many times, these imperfections appear on both surfaces in contact as 'peaks' and 'valleys' and they tend to interlock and resist motion (Fig. 17.2(a)). But friction exists even between very smooth surfaces, due to pressure and relative motion of the surfaces.

The effects of friction must be reduced in order to prevent

(a) loss of energy in overcoming frictional resistance;

(b) overheating of surfaces in contact, which can result in melting and fusing together of mating parts;

(c) damage to the surfaces in contact, due to wear.

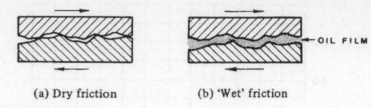

(a) Dry friction   (b) 'Wet' friction

Fig. 17.2 Sliding friction

Lubrication

The effect of dry sliding friction is reduced considerably by introducing a thin film of lubricant – oil or grease – which keeps the two sliding surfaces apart (Fig. 17.2(b)).

Oils are liquid in structure and are used mainly for the following:

(a) high speeds,

(b) high temperatures,

(c) as coolant to dissipate frictional heat,

(d) in splash and circulation systems of lubrication.

Greases are semisolid in structure and are used mainly for the following:

(a) low speeds,

(b) corrosion protection of metal bearings,

(c) prevention of foreign-matter contamination by partially sealing the bearings,

(d) applications where lubricant leakage must be prevented (food, paper, textile industries, etc.),

(e) long-life lubrication for sealed or shielded bearings.

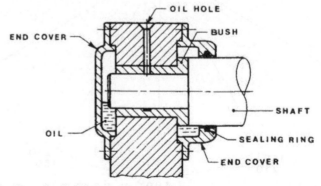

Fig. 17.3 A method of lubricating bearings

17.2 Plain bearings (Fig. 17.4)

Plain bearings consist of a supporting part, called the housing, and a mating part which may be a shaft, pivot, or thrust collar.

The ideal plain bearing should be hard, strong, and wear-resistant, with a soft overlay which can be easily deformed, to absorb sudden loads without fracturing, and in which foreign particles can be embedded, thus preventing rapid wear of the mating surfaces.

There are two classes of plain bearings: (a) direct-lined housings and (b) bushes. The materials used for them are discussed in sections 17.3 and 17.4.

Direct-lined housings

In this type of plain bearing, the housing is lined directly with bearing material by means of metallurgical bonding or keying. This construction is mostly confined to low-melting-point white metals attached to ferrous housings.

Applications: Cement-mill machinery, crushing plants, large crankshafts, car engines, etc.

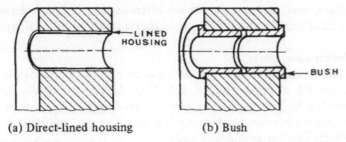

(a) Direct-lined housing   (b) Bush

Fig. 17.4 Plain bearings

Bushes

Bushes are hollow cylindrical pieces which are fitted into a housing to accommodate the mating part. When worn, they are removed and replaced.

To prevent the rotation of bushes in their housings, an interference fit is used. Alternatively, pins, screws, or various shapes of bush ends to fit in correspondingly shaped housing bores are utilised.

There are two types of bush: solid and lined. In both cases the bearing material might be a white metal, a copper alloy, or an aluminium alloy.

Solid bushes   These are made entirely of the bearing material.
Applications: Earth-moving equipment, gearboxes, crankshaft bearings, steering-gear linkages, diesel-engine small-ends, and general applications.

Lined bushes   In these, the bearing material is applied as a lining to a backing material.
Applications: Plant machinery, turbines, large diesel engines, marine gearboxes, etc.

Advantages of plain bearings

(a) They usually require only a small radial space.
(b) They are cheap to produce.
(c) They have vibration-damping properties.
(d) They are noiseless in operation.
(e) They can be easily machined.
(f) They can cope with trapped foreign matter.

Disadvantages of plain bearings

(a) They require large supplies of lubricant.
(b) They are suitable only for relatively low temperatures and speeds.
(c) When starting from rest, the initial resistance to motion is much larger than the running resistance, due to the slow build up of the lubricant film around the bearing surface.

17.3 Bearing alloys

Materials suitable for plain bearings require special properties, as follows:

(a) *compressive strength*, to support loads imposed by components;
(b) *fatigue strength*, to withstand stresses due to forces which are applied repeatedly and which may vary in magnitude and direction;
(c) *corrosion resistance*, to resist corrosion due to oil oxidation products, water, or other contaminants;
(d) *ductility*, to be able to deform very slightly in order to absorb sudden loads without fracturing;
(e) *thermal conductivity*, to conduct heat generated away from bearing contact surfaces;
(f) *low coefficient of friction*, to ensure low friction forces;
(g) *embeddability*, so that all trapped foreign particles can sink in below the bearing surface, thus preventing rapid wear of the mating surfaces;
(h) *machinability*, for economical and easy bearing production;
(i) *bonding property*, to allow the bearing to be attached firmly to a backing material if required.

Bearing alloys have been produced with various combinations of the above properties to suit different applications. The metal on which the alloy is based forms a matrix whose properties are modified by the alloying elements. The harder and stronger the matrix, the higher is its fatigue strength; conversely, the softer it is the better is its embeddability. A soft matrix will accommodate some displacement due to loading and any slight misalignment of a mating part and will allow evenly distributed wear, but it tends to wear more rapidly.

White metals

These alloys, also known as Babbit metals, consist mainly of tin and/or lead with small amounts of antimony and copper. The addition of antimony improves the compressive strength and provides resistance to wear but introducers brittleness. By adding a small amount of copper the desirable ductility and toughness are restored. Lead is used mainly in the interests of cheapness.

White metals have low melting points and are therefore not suitable for high-temperature applications, but their embeddability is very good.

Tin-base alloys are preferable to lead-base alloys as
(a) they flow more readily when molten,
(b) they shrink less when solidifying,
(c) they are more ductile,
(d) they are more corrosion-resistant.
However, they are more expensive to produce than lead-base alloys.

A typical composition of a tin-base bearing alloy used for general work would be 86% tin, over 10% antimony, and over 3% copper.

Lead-base alloys are more liable to be corroded by products of oxidation of oil and have a tendency for abrasive wear. They are used for low speeds and light duties.

A typical composition of a lead-base bearing alloy used for low-duty applications would be over 63% lead, 20% tin, 15% antimony, and over 1% copper.

Copper-base alloys

These alloys are harder, stronger, and more resistant to wear than white metals.

Copper–lead alloys Where loads are too high for white metals and sizes of bearings are restricted, an alloy of about 70% copper and 30% lead may be used. About 2% of tin may be added to reduce hardening and brittleness. These types of bearing are used for high speeds, notably in diesel engines.

Lead bronzes These alloys have a bronze matrix consisting of copper and tin, in which the lead is distributed. The hardness and fatigue

strength will depend upon the tin content, which may be from 1% to 15%. The lead content may be up to 25%. These bearings have excellent casting properties, are easily machined, and are used for high-duty applications.

Tin bronzes These copper–tin alloys are hard and strong. They have high fatigue and compressive strengths, good resistance to wear, but poor embeddability. They are used for low speeds and high-duty applications.

Phosphor bronzes These alloys have a small phosphorus content which has a deoxidising effect. They are corrosion-resistant, and stronger than simple tin bronzes.

Gun metals These are copper–tin–zinc alloys with occasional small amounts of lead. They are wear- and corrosion-resistant and are used for marine fittings, valves, etc.

Brasses These copper–zinc alloys with very small amounts of aluminium, iron, and manganese are often used as bearings. Bearing brasses have usually about 60% of copper and about 40% of zinc.

Aluminium-base alloys

These alloys may contain up to 20% tin, to provide embeddability, with about 1% of copper and nickel. Their fatigue strength is similar or somewhat higher than that of copper–lead and copper–tin bronzes.

They are mainly used as bearings for diesel engine and motor-car crankshafts and for connecting rods.

Table 17.1 Relative properties of some bearing alloys

Alloy	Hardness	Embedding ability	Corrosion resistance	Fatigue resistance	Thermal conductivity
Tin-base Babbit	Extremely low	Very high	Very high	Low	Medium
Lead-base Babbit	Extremely low	Very high	Medium	Low	Low
Copper–lead alloy	Low	Medium	Low	Medium	Very high
Lead bronze	Low	Medium	Medium	High	Medium
Tin bronze	Low	Low	High	Very high	Medium
Aluminium-base alloy	Very low	Medium	Very high	Very high	High

17.4 Other bearing materials

Cast-iron bearings

A cast-iron bearing is usually simply a hole bored in a cast-iron part to accommodate a mating part. For grey cast-iron, adequate lubrication free from dust is essential.

These types of bearing are in general used for light duties.

Porous metal bearings

Very fine metal powders are partially compressed to a required shape and then sintered at a high temperature. The metal sponges so produced may be filled with a low-friction thermoplastic for dry running or be impregnated with oil. Porous metals may absorb up to one third of their own volume of lubricant and when running they give off the lubricant, thus wetting the bearing surfaces. When stationary, most of the lubricant is reabsorbed again.

Due to their low compressive strength, these bearings are used for low-duty applications, such as control linkages, door hinges, etc.

Non-metallic bearings

Thermoplastics Thermoplastics are suitable for injection moulding and extrusion as they can be softened and reshaped with the application of heat. These materials include PTFE (polytetrafluoroethylene), nylon, polystyrene, polypropylene, etc.

Applications include bushes for medical equipment, food-preparation equipment, textile machinery, pumps, etc. PTFE, with its high thermal resistance and load-carrying capacity, is used for chemical pumps, railway-point pivot bushes, conveyor bushes, etc.

Thermosetting plastics Thermosets are chemically changed when heated – they become rigid, and this change cannot be reversed. For use as bearings these materials are always reinforced by the addition of asbestos, silicone resins, carbon, or metals.

These types of bearing are used for hot-duct supports, jet-pipe supports, cranes, vibratory rollers, pump bushes, etc.

Carbon (graphite) Graphite bearings have a very high resistance to elevated temperatures, need no lubrication, have a polishing effect on the mating parts, and can run in fluids which attack other bearing materials.

Carbon bearings are used for food and textile machinery, furnace and boiler equipment, chemical agitators, bottle-washing plant, trolley wheels, etc.

Rubber Rubber bearings are easily deflected, thus reducing stresses and damping vibrations. They are usually used for pumping purposes with water as the lubricant. Rusting mating surfaces should be avoided, as rough surfaces will damage the rubber.

Wooden bearings Lignum vitae is one of the hardest and densest of all woods. It is very strong in compression, can resist the action of certain chemicals, and is used for bearings in food- and chemical-processing machinery, etc.

Hard maple when impregnated with oil is used for textile machinery, loose pulleys, etc.

Jewel bearings are used for instruments, clock mechanisms, etc.

17.5 Ball and roller rolling bearings

In ball and roller bearings, the sliding friction of plain bearings is replaced by a much lower rolling friction.

Also, when starting plain bearings from rest, their initial resistance to motion is considerably larger than the running resistance after the lubricant film has been built up around the bearing surface. However, when starting ball and roller bearings from rest, their initial resistance to motion is only slightly more than their resistance to continued running. These bearings are therefore used for devices which are subject to frequent starting and stopping.

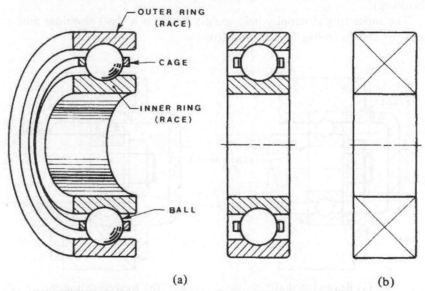

Fig. 17.5 Rolling-bearing parts (a) sectional view of ball bearing (b) conventional representation of a bearing

149

Bearing parts (Fig. 17.5)

Ball and roller bearings consist of

(a) an *inner ring*, which fits on the shaft;

(b) an *outer ring*, which fits inside the housing;

(c) *balls* or *rollers*, which provide a rolling action between the rings;

(d) a *cage*, which separates adjacent ball surfaces, which rotate in opposite relative directions, and prevents sliding friction between them.

Note: Rings sometimes are called 'races'.

Bearing materials

The materials used for rolling bearings are high-carbon chromium steels which are very hard and resistant to wear. Surfaces in contact must be highly polished to reduce wear and to provide smooth rolling movement without any sliding.

The cages are made of low-carbon steels, bronzes, or brasses, though for high-temperature applications case-hardened and stainless steels are used.

Fits

The inner ring of a journal bearing must have an interference fit on a revolving shaft, to prevent creep. (Creep is the slow rotary movement of the inner ring relative to the rotating shaft or, alternatively, the slow rotary movement of the outer ring relative to the rotating housing).

The inner ring is usually held axially between a shaft shoulder and a nut, as shown in Fig 17.6(a) and (b).

The outer ring is assembled inside the stationary housing, with a sliding fit to permit correct axial position without preload. Often the outer ring is not held axially – a clearance is introduced in order to allow for inaccurate machining and for relative expansion or contraction of the shaft and the housing due to temperature changes, as shown in Fig. 17.6(a).

For revolving parts on the shaft, such as wheels, pulleys, etc., the outer rings have an interference fit inside the housing of these wheels, whereas the inner rings have a sliding fit on the stationary shaft, as shown in Fig. 17.6(b).

General rule: Rotating rings require interference fits; stationary rings require sliding fits.

17.6 Types of rolling bearing

Single-row deep-groove ball bearings (Fig. 17.7(a))

These bearings incorporate a deep hardened raceway or track which makes them suitable for radial and axial loads in either direction, providing the radial loads are greater than the axial loads.

These bearings are self-contained units – they can be handled and assembled as a single component. Also, they may be of the prelubricated types, having integral seals or shields which retain the grease within the bearing and prevent foreign matter entering it (Fig. 17.11).

Applications These bearings can be used as locating bearings for high-speed spindles, motorcycle engines, electric motors, circular saws, turbine shafts, wood cutters, gearboxes, pulleys, pumps, etc.

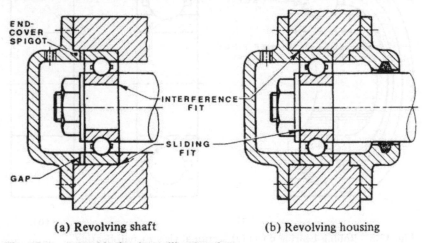

(a) Revolving shaft   (b) Revolving housing

Fig. 17.6 Assembly fits for rolling bearings

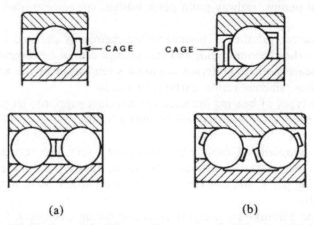

(a)   (b)

Fig. 17.7 (a) Single- and double-row deep-groove ball bearings (b) single- and double-row angular-contact ball bearings

Single-row angular-contact ball bearing (Fig. 17.7(b))
Outer rings and sometimes inner rings are machined with high and low shoulders to take one-directional thrust or combined radial and axial loading. To support thrust loads in either direction, these bearings can be mounted in opposing pairs, suitably adjusted to prevent end play or preloading due to over-adjustment.

Double-row angular-contact ball bearings can be used as alternatives to opposing pairs of single-row bearings, simplifying the mounting and saving space.
Applications   These bearings are used for high-speed spindles for boring, milling and drilling machines and for motorcycle engines, worm gears, hot-gas fans, radar aerials, etc.

Double-row self-aligning ball bearings (Fig. 17.8(a))
These bearings are designed for the cases where alignment of inner and outer rings cannot be assured during assembly or in service. The inner ring has two deep-groove raceways, whereas the outer ring has a single spherical raceway. This allows the inner and outer rings to be misaligned relative to each other.
Applications   These bearings should not be used for very heavy radial loads and they will not support any axial loads. Protective shields cannot be fitted. These bearings are used for gearboxes, air blowers, cutter shafts of planing machines, etc.

Single-row roller bearings (Fig. 17.8(b))
Roller bearings are usually detachable units. They have a greater load-carrying capacity than ball bearings of equivalent size as they make line contact rather than point contact with their rings. These bearings are not suitable for axial loading. A slight axial displacement is permissible between the inner and outer rings. These types of bearing are cheaper to manufacture than equivalent ball bearings.
Applications   These bearings are suitable for heavy and sudden loading, high speeds, and continuous service. They are used for vibrating motors, stone crushers, dredging machinery, ship propellers and rudder shafts, belt conveyors, locomotive axles, flywheels, crankshafts, presses, etc.

Tapered-roller bearings (Fig. 17.9(a))
These bearings are so designed that, when projected, construction lines corresponding to the surfaces of contact of rollers and rings will meet at an *apex* point on the bearing axis.

Like angular-contact ball bearings, these bearings will carry a combination of radial and single-direction thrust loads. The inner ring, called the cone, with a tapered roller and cage, is assembled as a complete unit; whereas the outer ring is detachable.

Two bearings can be mounted on a shaft, but they must be accurately adjusted axially to ensure the proper running clearance between the roller and the outer ring, called the cup.
Applications   These bearings are suitable for lathe spindles, bevel-gear transmissions, gearboxes for heavy trucks, car drives, car wheels, etc.

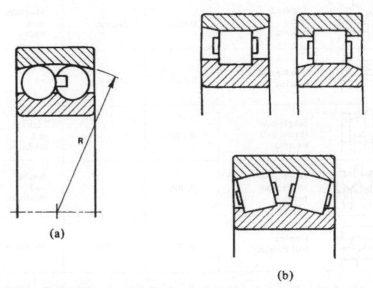

Fig. 17.8   (a) Double-row self-aligning ball bearing (b) single- and double-row roller bearings

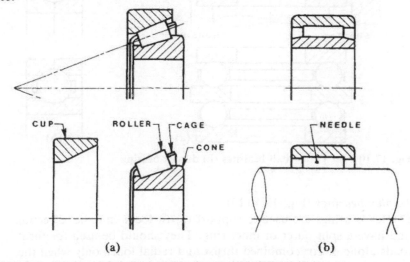

Fig. 17.9   (a) Tapered roller bearing (b) needle roller bearings

151

Needle-roller bearings (Fig. 17.9(b))

These bearings are fitted with small-diameter rollers and are used for radial loads at slow speeds and oscillating motion. They are especially suitable for restricted spaces and where the weight of components is critical, as in the aircraft industry.

Sometimes, due to space limitations, the inner ring is not used; instead, the shaft is hardened and ground and is used directly as an inner ring.

Applications   These bearings are used for aircraft applications, live tailstock centres, bench-drill spindles, light gearboxes, etc.

Thrust ball bearings (Fig. 17.10(a))

These bearings can only take thrust loads. They consist of two loose thrust rings grooved to accommodate the balls with their corresponding cage. One of the rings has a smaller bore than the other and engages on the shaft, while the other ring has a larger outer diameter which engages in the housing. A *double-thrust bearing* has three rings.

Applications   Thrust ball bearings are used for heavy axial loads and low speeds, and are suitable for thrust spindles, tailstock centres for heavy work, vertical shafts, pillars supporting heavy vertical loads, etc.

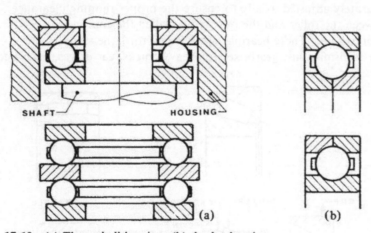

Fig. 17.10   (a) Thrust ball bearings (b) duplex bearing

Duplex bearings (Fig. 17.10(b))

Duplex bearings are used to support thrust loads in either direction and have a split outer or inner ring. They should be used for thrust loads alone or for combined thrust and radial loads only when the thrust load is very much greater than the radial one.

Table 17.2   Comparison of different types of rolling bearing

Type of bearing	Approximate coefficient of friction	Radial-load capacity	Axial-load capacity
Single-row deep-groove ball bearing (Double-row)	0.001	Light and medium (Heavy)	Light and medium (Medium)
Single-row angular-contact ball bearing	0.002	Medium	Medium
Double-row angular-contact ball bearing	0.002	Medium and heavy	Medium
Double-row self-aligning ball bearing	0.001	Medium	Light
Single-row roller bearing (Double-row)	0.001	Heavy (Very heavy)	–
Tapered-roller bearing	0.002	Heavy	Medium and heavy
Needle-roller bearing	0.003	Heavy	–
Single-row thrust ball bearing	0.001	–	Light and medium
Double-row thrust ball bearing	0.001	–	Heavy and medium
Duplex ball bearing	0.002	Very light	Medium

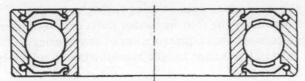

Fig. 17.11 Prelubricated sealed ball bearing

Prelubricated bearings (Fig. 17.11)

These bearings incorporate metal shields and/or seals, which are usually fastened to the outer ring. The close clearance between the seal and the inner ring retains the long-life grease within the bearing and prevents foreign matter entering from outside.

17.7 Design of rolling-bearing assemblies

Ball and roller bearings are normally mounted on a shaft, with the inner ring clamped against the shaft shoulder by means of a lock nut.

The outer ring is sometimes fixed endways between the spigot of the end cover and the housing shoulder, as shown in Fig. 17.12.

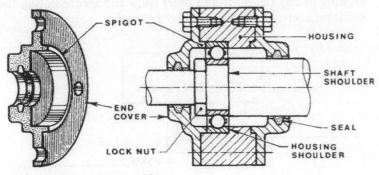

Fig. 17.12 Ball-bearing assembly

To ensure that a bearing operates properly it must be protected against foreign matter entering the housing, and at the same time the lubricant must be kept inside the bearing. This function is performed by the bearing seals or shields (Fig. 17.11), or by providing the end covers with contact or non-contact seals, as shown in Figs. 17.12 and 17.13.

The contact type of seal usually consists of a ring made of felt, leather, synthetic rubber, etc., fixed into the groove in the end cover and contacting the rotating shaft.

The non-contact type of seal is used when high temperatures and speeds are employed. A close clearance is provided by a series of grooves in the end cover and in the shaft or shaft-collar which are

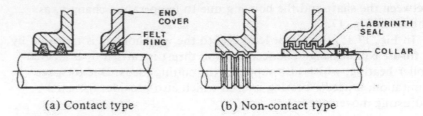

(a) Contact type   (b) Non-contact type

Fig. 17.13 Bearing seals

then filled with sealing grease. This method eliminates the friction and wear of the rubbing contact-type seals.

When a shaft is supported by two bearings, only one of the bearings should be fixed axially. The other bearing must provide axial adjusting movement by the ring sliding to accommodate the tolerances of position and to allow for relative differences in axial dimensions

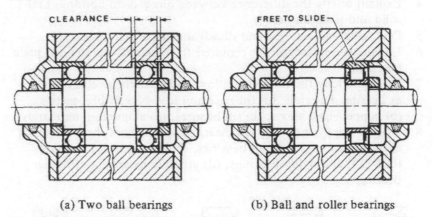

(a) Two ball bearings   (b) Ball and roller bearings

Fig. 17.14 Shaft supported by two bearings

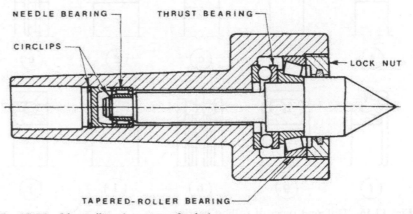

Fig. 17.15 Live tailstock centre of a lathe

between the shaft and the housing due to temperature changes, as shown in Fig. 17.14.

In Fig. 17.15, the thrust load due to the workholding is taken up by a thrust ball bearing. The feed-motion thrust is carried by a tapered-roller bearing, which also supports the cutting loads. Due to space limitation, a needle bearing is used which also provides the axial adjusting movement.

17.8 Test questions

1 Explain briefly the reasons for the use of bearings and say why the effects of friction on bearings must be reduced.
2 Discuss briefly the beneficial effect of using lubricants in bearings and state the relative advantages of using oil or grease.
3 What are the basic differences between journal and thrust bearings and between plain and rolling bearings?
4 Explain briefly the difference between direct-lined housings and solid and lined bushes.
5 Discuss the advantages and disadvantages of plain bearings.
6 List the special properties required for materials suitable for plain bearings.
7 Discuss briefly the following alloys and state their relative merits as bearing materials: (a) tin-base and lead-base Babbit metals, (b) copper–lead alloys, (c) tin bronzes, (d) aluminium-base alloys.
8 Discuss briefly the following bearings and suggest their industrial applications: (a) cast-iron bearings, (b) porous metal bearings, (c) PTFE and nylon bearings, (d) graphite bearings, (e) rubber bearings.

9 State the main advantages of using rolling bearings.
10 Sketch and name the four important parts of a roller bearing, and sketch a conventional representation of any bearing.
11 What materials are used for the manufacture of rolling-bearing parts?
12 Figure 17.16 shows different types of rolling bearing. Identify each of these bearings, stating the type of loading each will support and indicating suitable industrial applications.
13 State the types of fit employed when assembling rolling bearings for revolving shafts and stationary housings, and for stationary shafts and revolving housings. Give reasons for using these fits.
14 Name the rolling bearings which are *not* suitable to support the following: (a) thrust (axial) loads, (b) journal (radial) loads.
15 Name the rolling bearings which are suitable to support the following:
(a) a combination of thrust and journal loads in one direction only,
(b) a combination of thrust and journal loads in both directions.
16 Complete the sub-assembly drawing shown in Fig. 17.17 by locating axially the inner and outer rings and incorporating the lubricant seals. Tracing paper may be used.

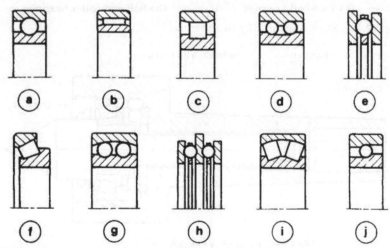

Fig. 17.16 Different types of rolling bearing

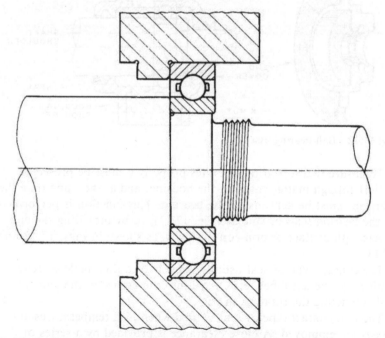

Fig. 17.17 Sub-assembly drawing of a rolling-bearing mounting

18 Basic engineering materials

There is a vast range of modern materials available to the engineering designer, who faces a formidable task when selecting the most suitable material for a particular component to perform a specific function, as there are so many different factors that might influence his choice. These factors include the design requirements, mass, loading, service life, climatic and chemical environment, reliability, machining requirements, manufacturing processes, properties of materials, and considerations of cost, quantity required, etc.

The properties of materials can be subdivided into:

(a) **Physical and chemical properties** These include melting point, thermal conductivity, electrical conductivity, density, coefficient of linear expansion, corrosion resistance, etc.

(b) **Mechanical properties** Some of these properties are discussed below.

Strength is the ability of a material to resist applied forces without fracture or permanent deformation. The applied forces may produce tension, compression, shear, or torsion. Impurities in the material, hot or cold working, heat treatment, and other factors can affect the strength of a material.

Elasticity is the ability of a material to return to its original shape after being deformed or strained due to stress resulting from the application of forces.

Ductility is the ability of a material to deform considerably under a tensile load before failure. Materials having this property can be formed into various shapes by bending, drawing, extruding, rolling, etc. Those possessing high ductility are gold, aluminium, silver, lead, and certain grades of steel and brass.

Brittleness is the property of a material of fracturing without much permanent distortion, especially when subjected to suddenly applied forces. Brittleness is the lack of ductility. Materials possessing brittleness are most cast irons, glass, and certain grades of steel, especially when the phosphorus content is too high.

Malleability is the ability of a material to be permanently deformed in compression without fracture; for example by hammering, forging, pressing, rolling, etc. This property is very similar to ductility. Materials possessing malleability are lead, tin, zinc, and gold.

Toughness is the ability of a material to resist fracture, especially under suddenly applied forces. It is the reverse of brittleness. Tough materials are capable of absorbing a large amount of impact energy without fracture and some can be repeatedly bent or twisted. Materials possessing toughness are wrought iron, aluminium, copper alloys, etc.

Hardness is the ability of a material to resist scratching, wear, abrasion, and indentation. The hardness of a material can be improved by alloying, or by a suitable heat treatment. Materials possessing high hardness include diamond, tungsten carbides, ceramics, heat-treated alloy cast irons, and some steels.

Machinability is the ability of a material to be machined with ease.

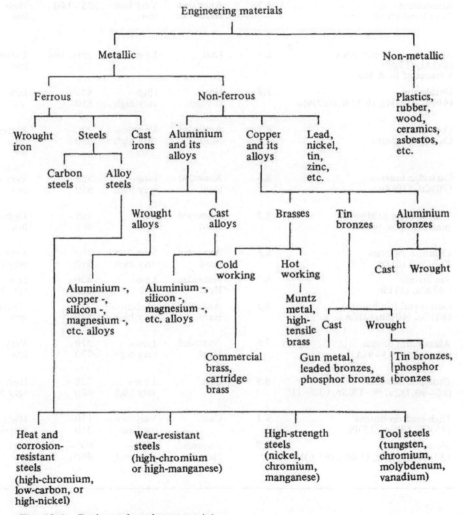

Fig. 18.1 Basic engineering materials

155

In general, engineering materials belong to two main groups – metals and non-metals – as shown in Fig. 18.1. Tables 18.1, 18.2 and 18.3 show the properties and indicate the typical uses of some engineering materials. These tables should serve as an approximate guide for selection of suitable materials for engineering components.

Table 18.1 Properties and uses of aluminium alloys and copper alloys. (These materials generally have pleasing appearance, with corrosion resisting properties).

Material and approximate composition (%)	Relative density	Condition	Elasticity	Tensile strength (MPa)	Ductility	Hardness	Machin-ability	Additional properties	Typical uses
Aluminium Al (almost pure)	2.7	Annealed Hard	Very low– low	\55–140	High– low	Very low– low	Fair	Good electricity & heat conductor; non-magnetic; malleable; can be forged & extruded; non-corrosive	Electrical cables, reflectors, cooking utensils, radiators, piping, building components, paints, etc.
Aluminium–silicon alloy (88)Al, (12)Si, + traces of Fe & Mn	2.7	Cast	Low	280–140	Extremely low	Very low	Fair	High castability and corrosion resistance; can be pressure die-cast	Light castings, aircraft & marine applications, radiators, crankcases, gearboxes, etc.
Duralumin (4)Cu, (0.8)Mg, (0.5)Si, (0.7)Mn	2.8	Heat-treated	High– very high	450– 550	Low	High	Good	Very high strength/weight ratio	General purposes, stressed aircraft components, structural components, etc.
Copper Cu (almost pure)	8.9	Annealed Hard	Very low– very high	220– 350	Extremely high – very low	Very low– medium	Poor	Good heat & electrical conductor; corrosion resistant; easily brazed; can be drawn & forged	Chemical industry, heating equipment, cooking utensils, tubing, roofing, boilers, etc.
Cartridge brass (70)Cu, (30)Zn	8.5	Annealed Hard	Low– very high	325– 650	Very high– low	Low– high	Poor	Non-magnetic; corrosion resistant; can be forged, drawn extruded	Cartridge, shells, jewellery, etc.
Yellow brass or Muntz metal (60)Cu, (40)Zn	8.3	Annealed Hard	Low– high	355– 465	High– low	Medium– very high	Fair	Non-magnetic; corrosion resistant; poor forming qualities	Structural plates, tubing, valve rods, hot forgings, etc.
Commercial brass (90)Cu, (10)Zn	8.8	Annealed Hard	Very low– very high	280– 510	Very high– very low	Low– high	Good	Can be drawn, bent, brazed, cold worked, welded, enamelled	Imitation jewellery, lipstick cases, clamps, etc.
Tin bronze (89)Cu, (11)Sn	8.7	Annealed Hard	Low	220– 310	Low– very low	Low– medium	Fair	Can be cast	Bearings, bushes, gears, piston rings, pump bodies, etc.
Gun-metal (tin bronze) (88)Cu, (10)Sn, (2)Zn	8.5	Annealed Hard	Medium– very high	270– 340	Medium– low	Low– medium	Fair	Good castability; resistant to salt-water corrosion	Bearings, steam valve bodies, marine castings, structural parts, etc.
Aluminium bronze (91–95)Cu, (5–9)Al	7.6	Annealed Hard	Low– very high	370– 770	Very high– low	Low– very high	Very poor	Corrosion & heat resistant; sea-water resistant; can be welded; difficult to machine	High wear and strength applications, marine hardware, etc.
Phosphor bronzes (86–90.7)Cu, (9–13)Sn, (0.3–1)P	8.9	Cast	Low– very high	220– 420	High– very low	Low– very high	Fair	Phosphorus improves tensile strength; wear and corrosion resistant	Bearings, bushes, valves, general sand castings, etc.
High-lead tin bronze (76)Cu, (9)Sn, (15)Pb	9.1	Cast	Very low– medium	170– 310	Medium– very low	Low– medium	Very good	Resistant to acid corrosion	General-purpose bearing and bushing alloy, wedges, etc.
Monel metal (30)Cu, (1.4)Fe, (1)Mn, (67.6)Ni	8.8	Annealed Hard	High– extremely high	600– 950	High– low	High– very high	Fair	Very high corrosion resistance; can be cast, forged, stamped, & drawn; heat resistant	Chemical engineering, propeller shafts, high-temperature valve seats, high-strength components, etc.

Aluminium (Al), 660   Carbon (C), 3550   Chromium (Cr), 1900   Lead (Pb), 327   Manganese (Mn), 1244   Nickel (Ni), 1453   Silicon (Si), 1412   Tin (Sn), 232

Antimony (Sb), 630   Copper (Cu), 1083   Iron (Fe), 1536   Magnesium (Mg), 650   Molybdenum (Mo), 2620   Phosphorus (P), 44   Sulphur (S), 113   Zinc (Zn), 420

Table 18.2 Properties and uses of irons, plain–carbon steels, and alloy steels. (These materials are generally cheap, widely available, and strong with a wide range of properties).

Material and approximate composition (%)	Relative density	Condition	Elasticity	Tensile strength (MPa)	Ductility	Hardness	Toughness	Machin-ability	Additional properties	Typical uses
Wrought iron Fe (almost pure)	7.8	Hot-rolled	Medium	340	Medium–high	Medium	High	Good	Malleable; shock resistant; can be bent, forged, hammer-welded	Chain links, ornamental work etc.
Grey cast irons (3–4)C, (1.2–2.8)Si, (0.5–1)Mn	7.2	Cast	Extremely low	170–350	Approx. zero	Very high	Very low	Good	Cheap; easily cast; corrosion & wear resistant; strong in compression; damps vibrations; 'self-lubricating'	Brackets, machine frames, pistons, cylinders, pipes, pulleys, gears, bearings, slides, etc.
Malleable irons (2–3)C, (0.1–0.5)P, (0.5–6)S, (1–5)Si, (0.4–21)Ni, (0.1–5)Cr	7.2	Cast Heat-treated	Medium	280–510	Very low–low	Medium	Low	Excellent–good	Increased ductility & malleability; shock & corrosion resistant; not weldable; can be cast and forged	Brake drums, levers, links, shafts, hinges, spanners, chains, wheels, vice bodies, etc.
Spheroidal-graphite (SG) irons Nodular irons	7.3	Cast	Very high	370–725	Medium	Very low–high	Medium	Fair–excellent	Added magnesium reduces graphite flakes to spheroids, increasing ductility, strength & shock resistance	Machine frames, pump bodies, pipes, crankshafts, hand tools, gears, dies, office equipment, etc.
Low-carbon steels (up to 0.25)C	7.85	Cannot be heat-treated	High	430–480	Medium	Medium	High	Fair–good	Most commonly used; cheap; magnetic; can be welded, forged, case-hardened	Lightly stressed parts, nails, car bodies, chains, rivets, wire, structural parts, etc.
Medium-carbon steels (0.25–0.6)C	7.85	Heat-treated	High–very high	480–620	Low–medium	High	Medium	Fair–poor	Weldable at lower carbon contents. With increased carbon contents brittleness is increased.	Axles, spindles, couplings, shafts, tubes, gears, forgings, rails, hand tools, dies, ropes, keys, etc.
High-carbon (tool) steels (0.6–1.5)C	7.85	Heat-treated	Very high–extremely high	620–820	Low	Very high	Very low	Very poor (anneal)	With increased carbon content brittleness is further increased & ductility is decreased	Hammers, chisels, screws, drills, taps, dies, blades, punches, knives, chisels, saws, razors, etc.
Nickel steels (0.1–4)C, (0.04–1.5)Mn, (1–5)Ni	7.8	Heat-treated	Medium–extremely high	310–700	Medium	High	Very high	Poor	Nickel improves strength and toughness	Axles, crankshafts, car parts, camshafts, gears, pins, pinions, etc.
Stainless steels (0.05–0.1)C, (0.8–1.5)Mn, (8.5–18)Ni, (12.5–18)Cr	7.9	Heat-treated	Medium–very high	650–900	Medium–high	Medium–very high	Medium–low	Good–fair	Over 12% chromium protects surfaces from corrosion 18/8 (Cr/Ni) steel is acid resistant.	Chemical plants, kitchen equipment, cutlery, springs, circlips, etc.
Low-alloy nickel-chrome steels (1–5)Ni, (0.6–1.5)Cr	7.8	Heat-treated	High–extremely high	930–1500	Low	Very high	High	Good	With heat treatment, a wide range of properties may be obtained.	Highly stressed parts: con-rods, shafts, gears, driving shafts, crankshafts, etc.
Manganese steels (0.35–1.2)C, (1.5–12.5)Mn	7.9	Heat-treated	Very high	700–850	Medium–high	Very high	High	Extreme-ly poor	Wear resistant; non-magnetic	Cutting tools, stone-crushing jaws, dredging equipment, press tools, railway crossings, etc.
Heat-resisting steel (0.1)C, (1.5)Si, (1)Mn, (19)Cr, (11)Ni	7.9	Air-cooled	High	690	High	Medium	High	Poor	Resistant to heat & thermal shock; 1000°C max. working temperature	Components exposed to high temperatures etc.

Table 18.3 Properties and uses of plastics. (These materials are very light, easily formed, sometimes transparent, corrosion resistant, with excellent electrical resisting properties).

Group	Compound	Relative density	Flamm-ability	Tensile strength (MPa)	Maximum working temperature (°C)	Chemical resistance	Relative cost	Additional properties	Typical uses
Thermoplastic materials. (These materials can be repeatedly softened and moulded by heating, then hardened by cooling.)									
Cellulosics	Cellulose nitrate (Celluloid)	1.37	Flammable	50	55	Fair	Low	Tough; dimensionally stable; low water absorption; inflammable; sensitive to heat	Handles, piano keys, toilet seats, fountain pens, spectacle frames, instrument labels, etc.
	Cellulose acetate (Tricel)	1.22–1.31	Slow burning	20–60	70	Fair	Low	Hard; stiff; strong; tough; transparent; surfaces of high gloss can be obtained.	Photographic film, artificial leather, lamp shades, toys, combs, cable covering, tool handles, etc.
Vinyls	Polyvinyl chloride (Rigid PVC) (Plasticized PVC)	1.34–1.40	Self-extinguishing	50	70–105	Good	Moderate	Rigid PVC is hard, tough, strong, stiff, abrasion resistant, Plasticized PVC is more flexible.	Pipes, bottles, chemical plant, lighting fittings, curtain rails, cable covers, toys, balls, gloves, etc.
	Polypropylene	0.9	Slow burning	35	100	Very good	Moderate	Strong; stiff; very light; good temperature resistance & electrical insulation properties	Pipes & fittings, bottles, crates, cable insulation, tanks, cabinets for radios, shoe heels, pumps, etc.
	Polystyrene	1.05–1.08	Slow burning	50	100	Good	Low	Rather brittle; transparent; Toughening with rubber improves impact & heat resistance	Vending machine cups, housings for cleaners & cameras, radio cabinets, furniture, toys, etc.
Fluorocarbons	Polytetrafluoroethylene (PTFE, 'Teflon', etc.)	2.13–2.19	Non-flammable	21	260	Excellent	Very high	Low coefficient of friction; tough; heat resistant; weather & corrosion resistant; can be machined	Bearings, gaskets, valves, chemical plant, electrical-insulation tapes, non-stick coatings for frying pans
Polyamides	Nylon 66	1.14	Self-extinguishing	70	150	Good	High	Stiff; strong; tough; abrasion resistant; high degree of rigidity; low-friction property	Bearings, gears, cams, pulleys, combs, bristles for brushes, ropes, fishing lines, raincoats, containers
Acrylics	Polymethylmethacrylate ('Perspex', 'Plexiglas', etc.)	1.19	Slow burning	55–80	93	Good	Moderate	Completely transparent; strong; stiff; shatter & weather resistant; will not discolour; can be decorated	Lenses, aircraft glazing, windows, roof lighting, sinks, baths, knobs, telephones, dentures, machine guards
Thermosetting materials (These materials undergo a chemical change when moulded: they become permanently rigid and incapable of being softened again.)									
Phenolics	Phenol formaldehyde ('Bakelite', etc.)	1.35–1.5	Very slow burning	50–60	120	Very good	Low	Brittle; heavy; hard; rigid; dark coloured; darken under influence of light; very popular thermoset	Vacuum cleaners, ashtrays, buttons, cameras, electrical equipment, dies, handles, gears, costume jewellery
	Melamine and Urea formaldehyde	1.4–1.55	Very slow burning to non-flammable	45–75	100	Good	Moderate	Available in light colours; good colour stability; brittle; heavy; hard; rigid	Electrical equipment, handles, cups, plates, trays, radio cabinets, knobs, building panels, etc.
Epoxides	Epoxy resins ('Bakelite', 'Araldite', etc.)	1.12–1.19	Slow burning to self-extinguishing	60	170	Good	Moderate	Due to many constituents, epoxides can be liquids, solutions, pastes, or solids	Adhesives, surface coatings, flooring, electrical insulation glass-fibre laminates, furniture, etc.
High-pressure laminates	Laminates ('Tufnol', 'Formica', etc.)	1.15–1.75	Slow burning to self-extinguishing	12–80	230–300	Good	High–low	Layers of paper, fabric, etc. bonded under pressure with a resin. Extremely strong; readily punched & machined; wear resistant	(Paper l's) electrical insulation. (Fabric l's) gears, bearings, jigs, aircraft parts, press tools. (Decorative l's) table tops, trays.

19 Introduction to engineering design

19.1 Engineering design definition

Engineering design may be defined as the process of using creativity, experience, technical information and scientific principles in the solution of engineering problems.

These solutions lead to the manufacture of a final product which must perform pre-specified functions with the maximum economy at the highest possible efficiency whilst satisfying environmental requirements.

19.2 Design process

The designer's responsibility covers the whole design process, which consists of a sequence of decisions taken by him or her or by a design team. It starts from the initial need for a product, leads to the design and manufacture of that product, then continues through servicing, modifications of design or even a complete redesign.

It is important to approach the design procedure systematically in order to reduce the design error, the need for redesign, and delays, and make it possible for more imaginative designs to be produced.

The systematic design process may be divided into a series of stages, which are not to be considered as separate independent steps but interacting related steps.

Design is a continuous iterative process, where modifications may be made at any stage. The designer may go back to any stage and even modify the design specifications, if required by new ideas and new technical developments.

19.3 Methodology of design

The recognition stage deals with the understanding of the design problem and the decision to do something about it.

This problem may be introduced to the designer by a customer, by their own firm or by their own recognition of the need for a certain product.

The analysis stage spotlights every aspect of the problem and involves consideration of the factors and restrictions that influence the possible solutions.

The designer considers all the customer's requirements, decides which of these must be fulfilled, which are not practical owing to certain restrictions, which are unimportant and can be neglected and what additional requirements may be introduced.

After considering all these factors, the designer prepares their own specifications, which are the parameters that the design solution must satisfy. These specifications must include all the requirements that are desirable features of the design solution, such as function, cost, safety, reliability, strength, appearance, service life, size and weight, energy consumption, method of manufacture, choice of materials, chemical and climatic conditions and ease of assembly and servicing.

All the individual factors should have a relative importance rating allocated to them, in order that the designer has guidelines within which to make decisions. Incorrect design specifications may cause unnecessary features to be incorporated in the manufactured product.

A design problem is broken into a set of general design specifications. The designer should use as many sources of information as possible, having in mind that for some design problems a particular source of information may be more applicable than the others: previous design solutions, handbooks, British Standard publications, technical journals and publications, textbooks, professional institutions and experimental establishments, industrial specialists, patent specifications, etc.

At this stage the designer seeks further information, calculations are made or experiments are carried out to determine the performance of the preliminary design solutions.

Also at this stage a *feasibility study* is introduced whereby alternative solutions are analysed, in order to establish whether they can be accepted and realised. The proposed specifications and alternative solutions are checked to see whether it is technically possible to produce them, whether they will function, be economic and be acceptable to the public.

The synthesis stage puts together all the ideas to form possible complete solutions. All the design factors and their importance are considered and irrelevances are discarded. This is the stage where numerous creative thoughts and ideas are considered and recorded. Here the designer might prepare a number of rough design sketches for alternative solutions.

The evaluation stage involves estimation or measurement of the degree to which the solutions satisfy the requirements. Merit points are allocated to each solution according to the degree to which it satisfies the design factors. The solutions are then retained or rejected on the basis of the number of merit points each accumulates. This method enables the designer to arrive at the *optimum* solution by methodical evaluation. Design can also be evaluated by relying on a designer's experience, prediction, experimenting, and use of models and prototypes.

At this stage the designer accepts their major professional responsibility.

The implementation stage presents the optimum solutions in the form of engineering drawings, pictorial sketches, models and photographs.

To complete the design and draughting phase a set of working drawings is prepared to enable the manufacture to be implemented.

The manufacturing stage does not signal the end of the design responsibility, which is integrated with the manufacturing process, prototype trials and customer trials. Further feedback information is considered carefully by the designer at all times and may entail modifications to or even a redesign of the already manufactured product.

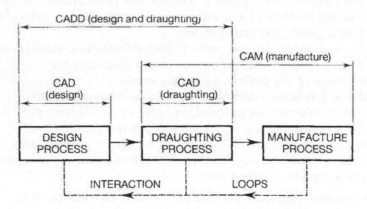

Fig. 19.1

19.4 Computer aided processes
The relationship between the computer aided processes of design, draughting and manufacture is shown in Fig. 19.1.

19.5 Summary of the systematic design process
The interactions between the various stages of the systematic design process are shown in Fig. 19.2.

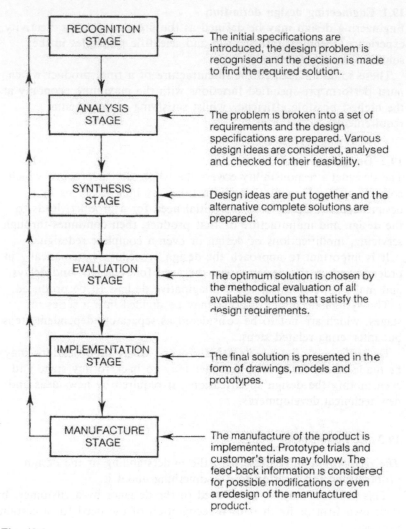

Fig. 19.2

160

Self-assessment questions

1 A rod EF is free to move in such a way that E is always in contact with a line OY and F is always in contact with another line OX.

Select from the diagrams shown below the locus of the mid-point P.

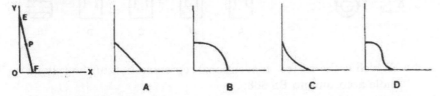

2 Using BS 4500 A, select the correct maximum and minimum size limits to give the $\varnothing$18 H8/f7 fit.
   (a) Maximum limits
   (b) Minimum limits
   A 18.033/17.980
   B 18.027/17.984
   C 18.000/17.966
   D 18.000/17.959.

3 Match the thread features shown below with their correct names:
   (a) crest
   (b) flank
   (c) thread angle
   (d) pitch.

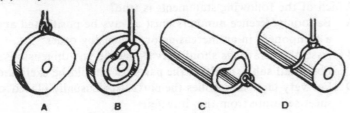

4 Select from four types of cam shown below:
   (a) a disc cam
   (b) a cylindrical cam.

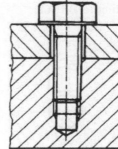

5 Four diagrams shown below indicate the different motions for a cam-follower rise. Which one shows a uniform acceleration?

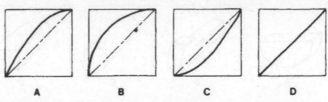

6 Match the tolerance symbols shown below with their correct names:
   (a) roundness
   (b) cylindricity
   (c) concentricity.

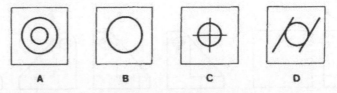

7 Of the assemblies shown below, which uses the correct conventional representation according to BS 308:

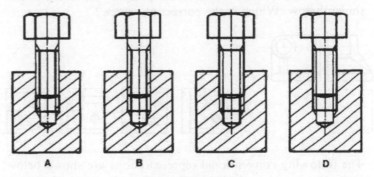

8 The two components shown in section are joined together by means of:
   A a bolt
   B a stud
   C a screw.

161

9 Which of the keys shown below is most suitable for a keyway in a tapered part of a shaft?

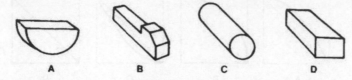

10 A front view, a plan view, and a space for an end view are shown below. Which is the correct end view?

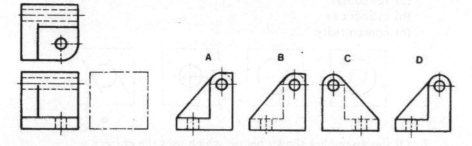

11 A front view, and end view, and a space for a plan view are shown below. Which is the correct plan view?

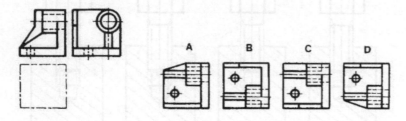

12 The following conventional representations are shown below: internal screw thread, diamond knurling, interrupted view, spur gear, and bearing. Which uses the correct conventional representation according to BS 308:

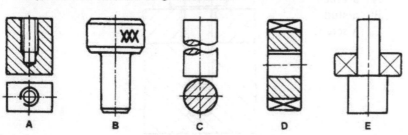

13 Which is the correct sectional view on XX below?

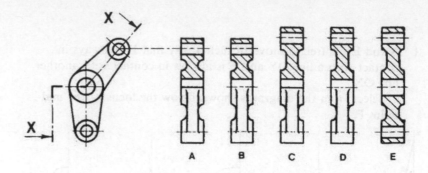

14 Which is the correct conventional representation of the splined hole according to BS 308:

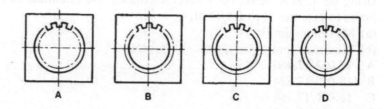

15 Each drawing of a component shown below is in orthographic projection and consists of a front view, an auxiliary view, and a projection symbol. Which drawing is correct?

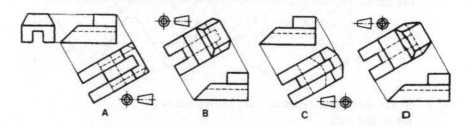

16 Which of the following statements is true?
A Balloon reference numbers must always be positioned around a component in an increasing or decreasing order.
B Assembly drawings should never include any dimensions.
C For small sub-assemblies the parts list should be excluded.
D For very large assemblies the parts list is usually placed on a sheet separate from the drawing.

Selected solutions

Solutions to the descriptive questions can readily be found from the text; solutions to a selection of other questions are given here, though in some cases alternative correct solutions are possible.

All diagrams are drawn to a reduced scale, and, for simplicity, the fillets on certain drawings have been omitted.

Page 13, Q 20 (a) 9, (b) 8, (c) 2, (d) 13, (e) 3, (f) 10, (g) 4, (h) 14, (i) 11, (j) 5, (k) 6, (l) 1, (m) 7, (n) 12

Page 13, Q 21

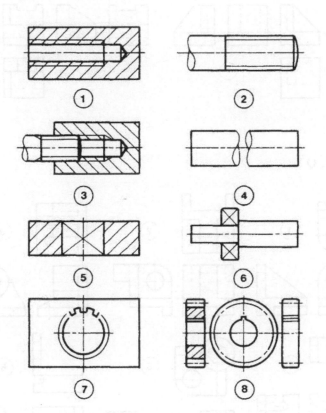

Page 21, Q 1

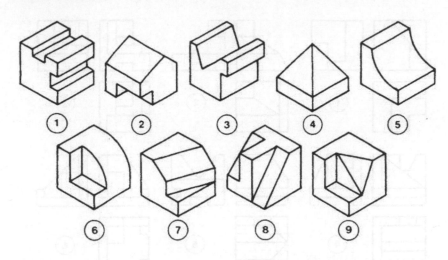

Page 22, Q 2

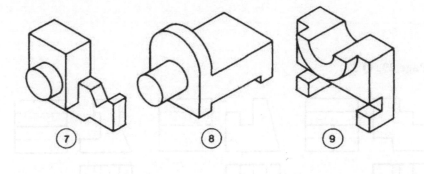

Page 25, Q 1

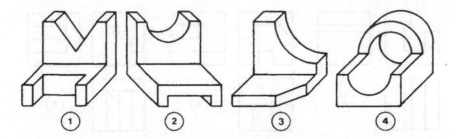

Page 28, Q 2

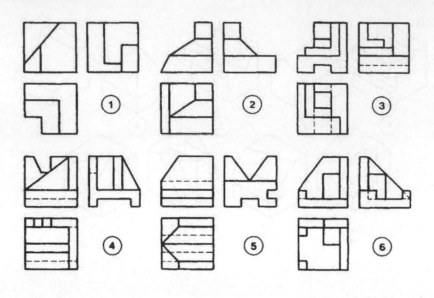

Page 29, Q 3

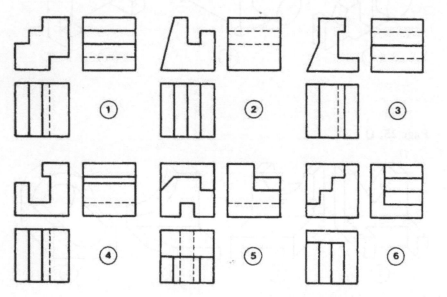

Page 32, Q 1

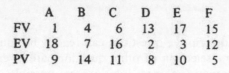

	A	B	C	D	E	F
FV	1	4	6	13	17	15
EV	18	7	16	2	3	12
PV	9	14	11	8	10	5

Page 33, Q 2

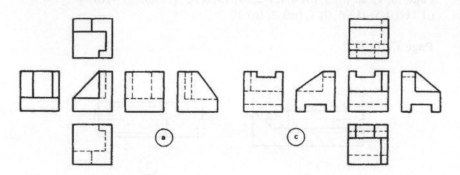

Page 33, Q 3

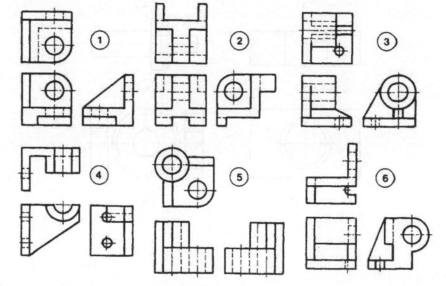

Page 34, Q 4

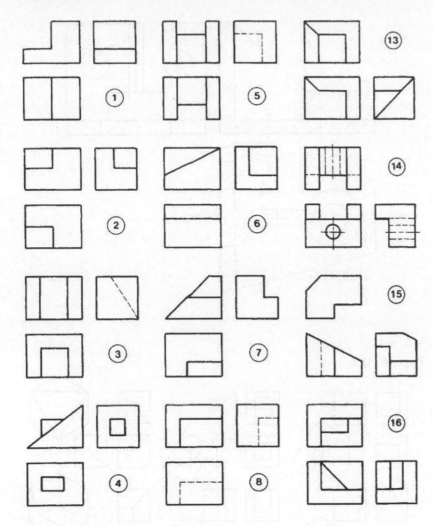

Page 35, Q 5

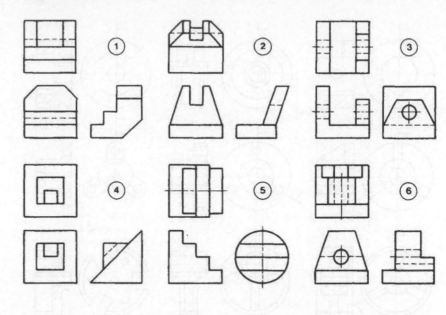

Page 36, Q 6 (a) B, (b) D, (c) E

Page 36, Q 7 (a) A, (b) D

Page 37, Q 9 (1) A, (2) B, (3) C, (4) D, (5) C, (6) A, (7) A, (8) C, (9) D

Page 41, Q 1

Page 42, Q 2

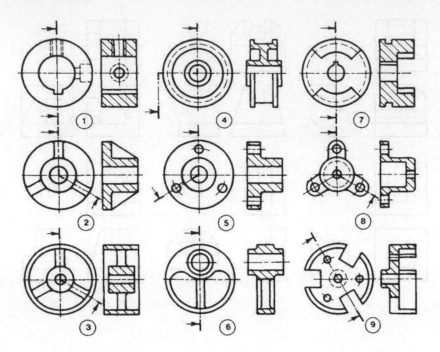

Page 43, Q 3

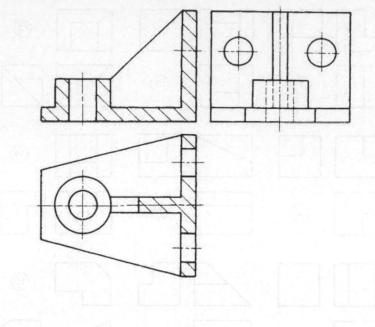

Page 45, Q 1

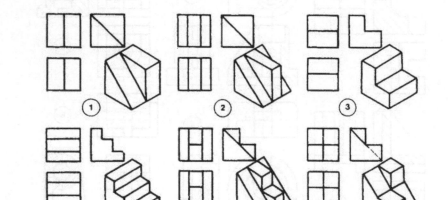

Page 46, Q 2

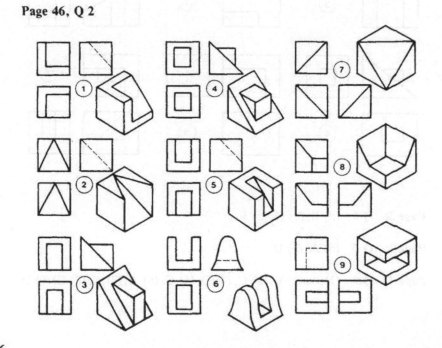

167

Page 76, Q 18

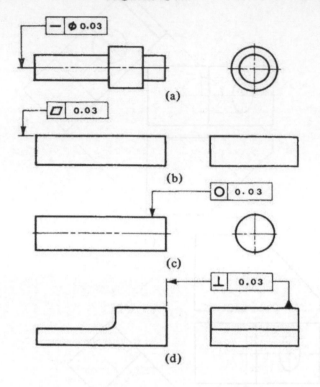

(a)

(b)

(c)

(d)

Page 80, Q 3 (a) T, (b) T, (c) F, (d) F, (e) F, (f) T, (g) F

Page 81, Q 5

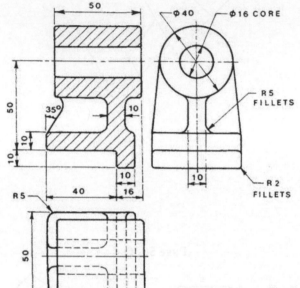

Page 82, Q 18   9.96, 9.22

Page 82, Q 19

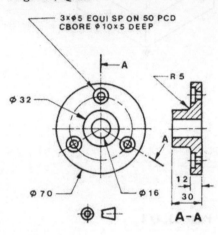

3×φ5 EQUI SP ON 50 PCD
CBORE φ10×5 DEEP

A-A

Page 82, Q 20

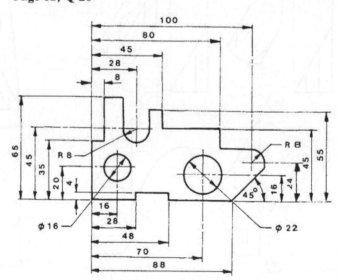

Page 80, Q 4

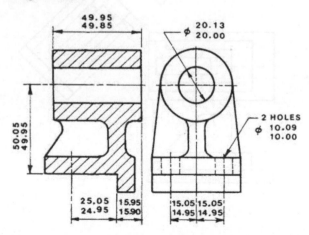

Page 82, Q 14

(a) hole   50,160   shaft   49.870
            50.000   49.710
(b) hole 100.035   shaft 100.045
           100.000   100.023
(c) hole 150.040   shaft 150.125
           150.000   150.100

168

Page 83, Q 21

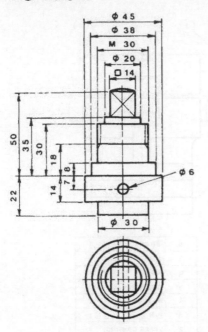

ϕ 45
ϕ 38
M 30
ϕ 20
□ 14
ϕ 6
50
35
30
18
14 8 7
22
ϕ 30

Page 86, Q 30

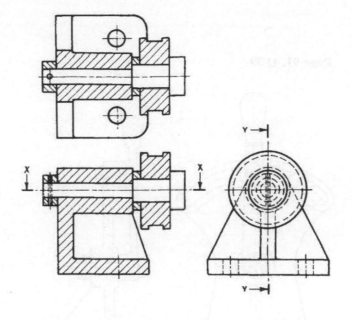

45 45
3 x ϕ 16
ϕ 50
95
11
20 20
ϕ 80
ϕ 120°
100
80
50
10

Page 87, Q 31

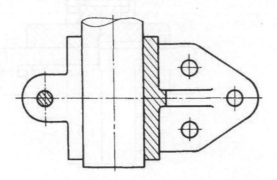

Page 83, Q 23 1

Page 88, Q 33

Y
X X
Y

169

Page 90, Q 37

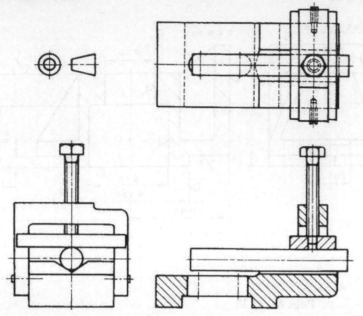

Page 90, Q 38

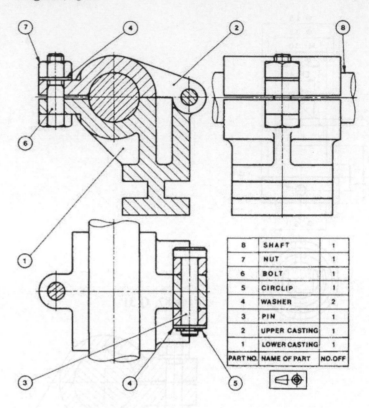

PART NO.	NAME OF PART	NO.OFF
8	SHAFT	1
7	NUT	1
6	BOLT	1
5	CIRCLIP	1
4	WASHER	2
3	PIN	1
2	UPPER CASTING	1
1	LOWER CASTING	1

Page 91, Q 39

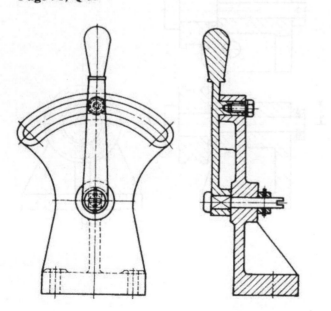

Page 93, Q 43

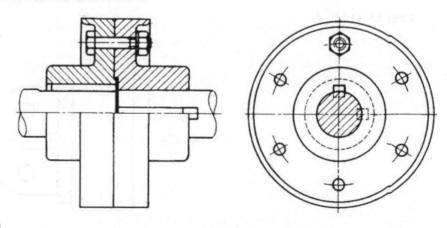

Page 93, Q 44

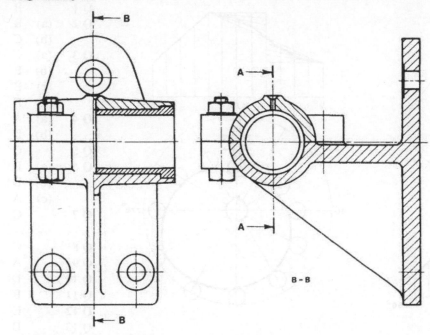

Page 94, Q 45

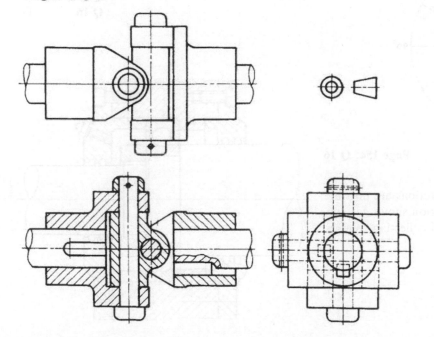

Page 95, Q 47

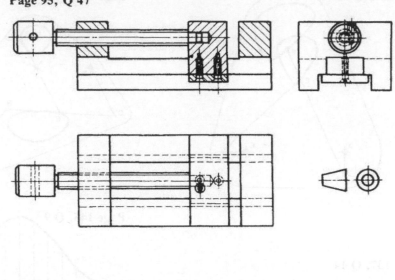

Page 96, Q 48

See p. 79, Fig. 7.1

Page 97, Q 49

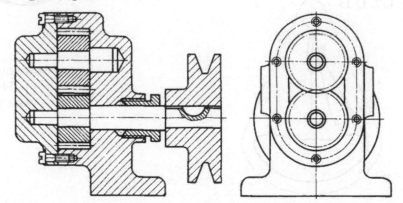

Page 137, Q 13

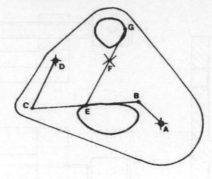

Page 137, Q 17

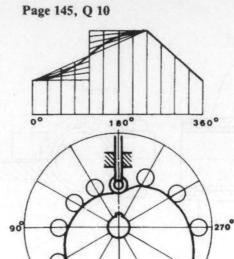

Page 145, Q 10

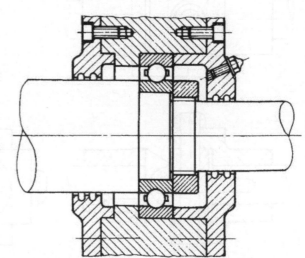

Page 137, Q 14

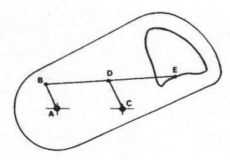

Page 145, Q 9

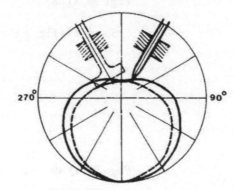

Page 137, Q 15

Page 154, Q 16

Many correct solutions are possible.
This solution shows gap-type
seals with oil-collecting recesses.

Index

T - #0239 - 071024 - C0 - 216/276/10 - PB - 9780415502900 - Gloss Lamination